SCHAUM'S OUTLINE OF

THEORY AND PROBLEMS

OF

ELECTRIC MACHINES
and
ELECTROMECHANICS

Second Edition

•

SYED A. NASAR, Ph.D.
Professor of Electrical Engineering
University of Kentucky

•

SCHAUM'S OUTLINE SERIES
McGRAW-HILL
New York   St. Louis   San Francisco   Auckland   Bogotá   Caracus   Lisbon
London   Madrid   Mexico City   Milan   Montreal   New Delhi
San Juan   Singapore   Sydney   Tokyo   Toronto

SYED A. NASAR is Professor of Electrical Engineering at the University of Kentucky. Having earned the Ph.D. degree from the University of California at Berkeley, he has been involved in teaching, research, and consulting in electric machines for over 40 years. He is the author, or co-author, of five other books and over 100 technical papers.

Schaum's Outline of Theory and Problems of

ELECTRIC MACHINES AND ELECTROMECHANICS

12 13 14 15 16 17 18 19 20 CUS CUS 0

ISBN 978-0-07-183738-5

Sponsoring Editor: Barbara Gilson
Production Supervisor: Tina Cameron
Editing Supervisor: Maureen B. Walker

Library of Congress Cataloging-in-Publication Data

Nasar, S. A.
    Schaum's outline of theory and problems of electric machines and
electromechanics / Syed A. Nasar. — 2nd ed.
       p. cm. — (Schaum's outline series)
    Includes index.
    ISBN 0-07-045994-0 (pbk.)
    1. Electric machinery. 2. Electric machinery—Problems,
exercises, etc. 3. Electric machinery—Outlines, syllabi, etc.
I. Title. II. Series.
TK2181.N38 1997
621.31'042'076—dc21
                                         97-23529
                                           CIP

McGraw-Hill

A Division of The McGraw·Hill Companies

Preface

A course in electric machines and electromechanics is required in the undergraduate electrical engineering curriculum in most engineering schools. This book is aimed to supplement the usual textbook for such a course. It will also serve as a refresher for those who have already had a course in electric machines or as a primer for solo study of the field. In each chapter a brief review of pertinent topics is given, along with a summary of the governing equations. In some cases, derivations are included as solved problems.

The range of topics covered is fairly broad. Beginning with a study of simple dc magnetic circuits, the book ends with a chapter on electronic control of dc and ac motors. It is hoped that the presentation of over 400 solved or answered problems covering the entire range of subject matter will provide the reader with a better insight and a better feeling for magnitudes.

In this second edition, the theme of the first edition is retained. Major additions and deletions include an expanded discussion of the development of equivalent circuits of transformers and the addition of a section on instrument transformers in Chapter 2; and the addition of a section on energy-efficient induction motors. In the revision, Chapter 7 has undergone major changes: Sections on linear induction motors, electromagnetic pumps, and homopolar machines are deleted. The chapter focuses on small electric motors. Thus, sections on starting of single-phase induction motors, permanent magnet motors, and hysteresis motors are added. In Chapter 8, the section on power semiconductors is completely rewritten. Finally, new problems are added to every chapter.

S. A. Nasar

Contents

Chapter 1

Magnetic Circuits

1.1 INTRODUCTION AND BASIC CONCEPTS

Electric machines and electromechanical devices are made up of coupled electric and magnetic circuits. By a *magnetic circuit* we mean a path for magnetic flux, just as an electric circuit provides a path for the flow of electric current. Sources of magnetic fluxes are electric currents and permanent magnets. In electric machines, current-carrying conductors interact with magnetic fields (themselves arising from electric currents in conductors or from permanent magnets), resulting in electromechanical energy conversion.

Consider a conductor of length *l* placed between the poles of a magnet. Let the conductor carry a current *I* and be at right angles to the magnetic flux lines, as shown in Fig. 1-1. It is found experimentally that the conductor experiences a force **F**, the direction of which is shown in Fig. 1-1 and the magnitude of which is given by

$$F = BIl \tag{1.1}$$

Here, *B* is the magnitude of the *magnetic flux density* **B**, whose direction is given by the flux lines. The SI unit of **B** or *B* is the *tesla* (T). (Another, equivalent unit will be introduced shortly.) Notice from (*1.1*) that *B* could be defined as the force per unit current moment. Equation (*1.1*) is a statement of *Ampere's law*; the more general statement, which holds for an arbitrary orientation of the conductor with respect to the flux lines, is

$$\mathbf{F} = I\mathbf{l} \times \mathbf{B} \tag{1.2}$$

where **l** is a vector of magnitude *l* in the direction of the current. Again the force is at right angles to both the conductor and the magnetic field (Fig. 1-2). Ampere's law, (*1.1*) or (*1.2*), providing as it does for the development of force, or torque, underlies the operation of electric motors.

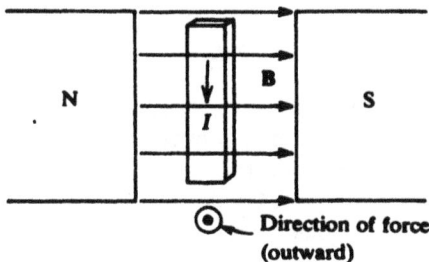

Fig. 1-1

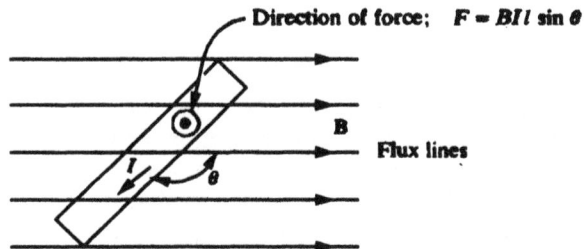

Fig. 1-2

The magnetic flux, ϕ, through a given (open or closed) surface is the flux of **B** through that surface; i.e.,

$$\phi = \int_s \mathbf{B} \cdot d\mathbf{S} = \int_s \mathbf{B} \cdot \mathbf{n} \, dS \tag{1.3}$$

where **n** is the unit outward normal to the elementary area *dS* of the surface (Fig. 1-3). In case **B** is constant in magnitude and everywhere perpendicular to the surface, of area *A*, (*1.3*) reduces to

$$\phi = BA \qquad (1.4)$$

from which

$$B = \frac{\phi}{A} \qquad (1.5)$$

The SI unit of magnetic flux is the *weber* (Wb). We see from (*1.5*) that B or $\mathbf{B}$ may be expressed in Wb/m^2, i.e., 1 T = 1 Wb/m^2.

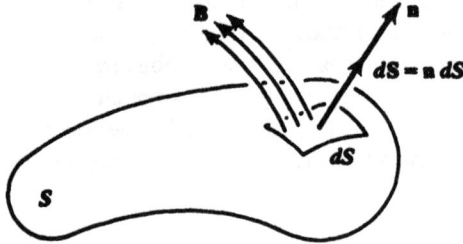

Fig. 1-3

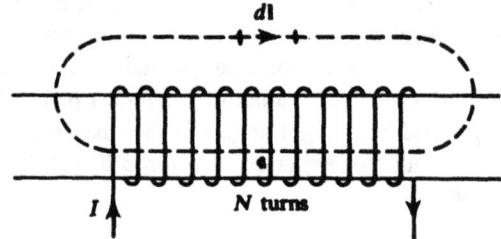

Fig. 1-4

The mutual relationship between an electric current and a magnetic field is given by *Ampere's circuital law*, one form of which is

$$\oint \mathbf{H} \cdot d\mathbf{l} = I \qquad (1.6a)$$

where $\mathbf{H}$ is defined as the *magnetic field intensity* (in A/m) due to the current I. According to (*1.6a*), the integral of the tangential component of $\mathbf{H}$ around a closed path is equal to the current enclosed by the path. When the closed path is threaded by the current N times, as in Fig. 1-4, (*1.6a*) becomes

$$\oint \mathbf{H} \cdot d\mathbf{l} = NI \equiv \mathscr{F} \qquad (1.6b)$$

in which $\mathscr{F}$ (or NI) is known as the *magnetomotive force* (abbreviated mmf). Strictly speaking, $\mathscr{F}$ has the same units, amperes, as I. However, in this book we shall follow the common convention of citing $\mathscr{F}$ in *ampere turns* (At); that is, we shall regard N as carrying a dimensionless unit, the *turn*.

Magnetic flux, magnetic flux density, magnetomotive force, and (see Section 1.2) permeability are the basic quantities pertinent to the evaluation of the performance of magnetic circuits. The flux, ϕ, and the mmf, $\mathscr{F}$, are related to each other by

$$\phi = \frac{\mathscr{F}}{\mathfrak{R}} \qquad (1.7)$$

where $\mathfrak{R}$ is known as the reluctance of the magnetic circuit.

1.2 PERMEABILITY AND SATURATION

In an isotropic, material medium, $\mathbf{H}$, which is determined by moving charges (currents) only, and $\mathbf{B}$, which depends also on the properties of the medium, are related by

$$\mathbf{B} = \mu\mathbf{H} \qquad (1.8)$$

where μ is defined as the *permeability* of the medium, measured in *henries per meter* (H/m). (For the *henry*, see Section 1.8.) For free space, (*1.8*) gives

$$B = \mu_0 H \qquad\qquad (1.9)$$

where μ_0, the permeability of free space, has the value $4\pi \times 10^{-7}$ H/m.

   The core material of an electric machine is generally ferromagnetic, and the variation of B with H is nonlinear, as shown by the typical *saturation curve* of Fig. 1-5(a). It is clear that the slope of the curve depends upon the operating flux density, as classified in regions I, II, and III. This leads us to

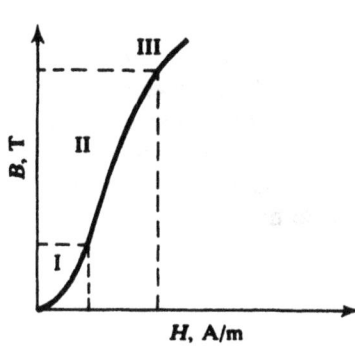

Fig. 1-5(a)

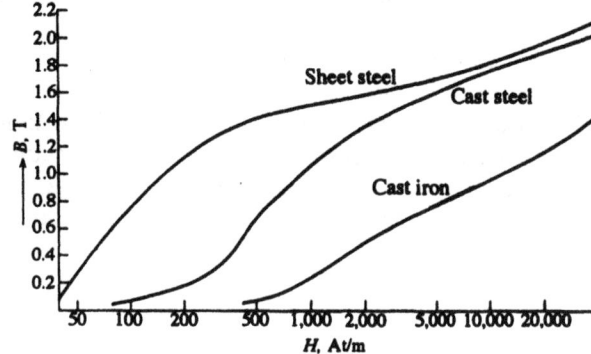

Fig. 1-5(b)

the concept of different types of permeabilities. We rewrite (*1.8*) as

$$B = \mu H = \mu_r \mu_0 H \qquad\qquad (1.10)$$

in which μ is termed permeability and $\mu_r = \mu/\mu_0$ is called *relative permeability* (which is dimensionless). Both μ and μ_r vary with H along the B-H curve. In the following, relative permeability is assumed; that is, the constant μ_0 is factored out. The slope of the B-H curve is called *differential permeability*;

$$\mu_d \equiv \frac{1}{\mu_0} \frac{dB}{dH} \qquad\qquad (1.11)$$

The *initial permeability* is defined as

$$\mu_i \equiv \frac{1}{\mu_0} \lim_{H \to 0} \frac{B}{H} \qquad\qquad (1.12)$$

The (relative) permeability in region I is approximately constant and equal to the initial permeability. In all three regions, the ratio of B to H at a point on the curve is known as *amplitude permeability*:

$$\mu_a \equiv \frac{1}{\mu_0} \frac{B}{H} \qquad\qquad (1.13)$$

Different ferromagnetic materials have different saturation curves, as shown in Fig. 1-5(b).

1.3 LAWS GOVERNING MAGNETIC CIRCUITS

   In some respects, a magnetic circuit is analogous to a dc resistive circuit; the similarity is summarized in Table 1-1.

**Table 1-1.   Analogy between a dc electric circuit
and a magnetic circuit**

Electric Circuit	Magnetic Circuit
Ohm's law, $I = V/R$	$\phi = \mathscr{F}/\mathfrak{R}$
resistance, $R = l/\sigma A$	reluctance, $\mathfrak{R} = l/\mu A$
current, I	flux, ϕ
voltage, V	mmf, $\mathscr{F}$
conductivity, σ	permeability, μ
conductance, G	permeance, $\mathscr{P}$

In the table, l is the length and A is the cross-sectional area of the path for the flow of current in the electric circuit, or for the flux in the magnetic circuit. In a magnetic circuit, however, l is the mean length of the flux path. Because ϕ is analogous to I and $\mathfrak{R}$ is analogous to R, the laws of resistors in series or parallel also hold for reluctances. The basic difference between electrical resistance, R, and magnetic reluctance, $\mathfrak{R}$, is that the former is associated with an energy loss (whose rate is I^2R), while the latter is not. Also, magnetic fluxes take leakage paths (Fig. 1-6), whereas electric currents normally do not.

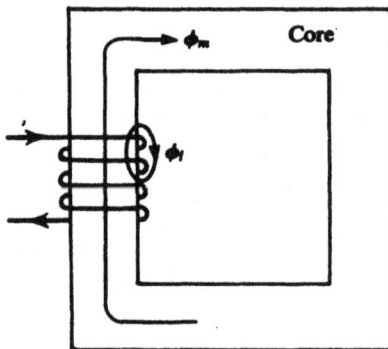

Fig. 1-6. Path of leakage flux, ϕ_l.

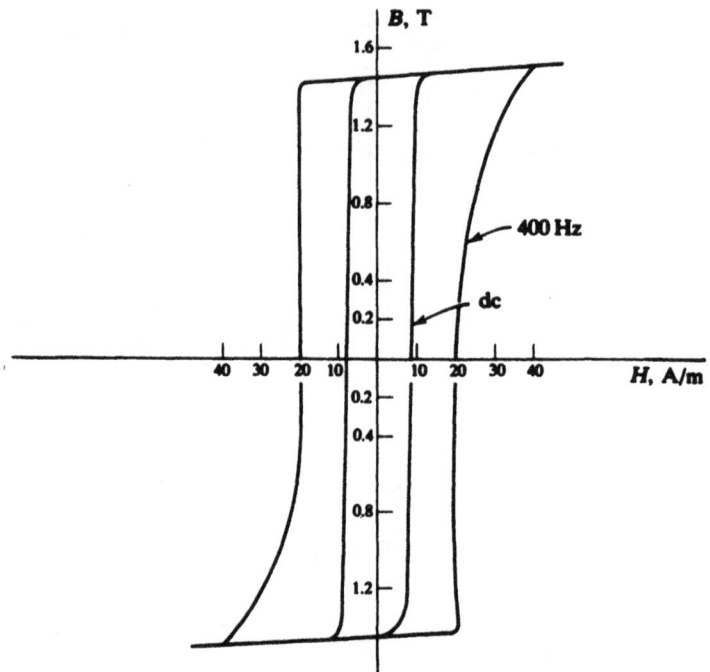

Fig. 1-7. Deltamax tape-wound core 0.002-in strip hysteresis loop.

1.4 AC OPERATION AND LOSSES

If the mmf is ac, then the B-H curve of Fig. 1-5 is replaced by the symmetrical *hysteresis loop* of Fig. 1-7. The area within the loop is proportional to the energy loss (as heat) per cycle; this energy loss is known as *hysteresis loss*.

Eddy currents induced in the core material (Fig. 1-8) constitute another feature of the operation of a magnetic circuit excited by a coil carrying an alternating current. The losses due to hysteresis and eddy currents—collectively known as *core losses* or *iron losses*—are approximately given by the following expressions:

$$\text{eddy-current loss:} \qquad P_e = K_e f^2 B_m^2 t^2 \text{ (W/kg)} \qquad\qquad (1.14)$$

$$\text{hysteresis loss:} \qquad P_h = K_h f B_m^{1.5 \text{ to } 2.5} \text{ (W/kg)} \qquad (1.15)$$

In (1.14) and (1.15), B_m is the maximum flux density, f is the ac frequency, K_e is a constant depending upon the material conductivity and thickness, and K_h is another proportionality constant. In addition, in (1.14), t is the lamination thickness (See Sec. 1.5).

1.5 STACKING FACTOR

To reduce eddy-current loss, a core may be constructed of laminations, or thin sheets, with very thin layers of insulation alternating with the laminations. The laminations are oriented parallel to the direction of flux, as shown in Fig. 1-8(b). Eddy-current loss is approximately proportional to the square of lamination thickness, which varies from about 0.05 to 0.5 mm in most electric machines. Laminating a core increases its volume. The ratio of the volume actually occupied by the magnetic material to the total volume of the core is known as the *stacking factor*; Table 1-2 gives some values.

Table 1-2

Lamination Thickness, mm	Stacking Factor
0.0127	0.50
0.0254	0.75
0.0508	0.85
0.10 to 0.25	0.90
0.27 to 0.36	0.95

(a) Unlaminated   (b) Laminated

Fig. 1-8

Because hysteresis loss is proportional to the area of the hysteresis loop, the core of a machine is made of "good" quality electrical steel which has a narrow hysteresis loop. Tape wound cores also have lower losses. Magnetic properties of some core materials are given in Appendix C.

1.6 FRINGING

Fringing results from flux lines appearing along the sides and edges of magnetic members separated by air, as shown in Fig. 1-9; the effect increases with the area of the airgap. Fringing increases with the length of the airgap.

1.7 ENERGY STORED IN A MAGNETIC FIELD

The potential energy, W_m, stored in a magnetic field within a given volume, v, is defined by the volume integral

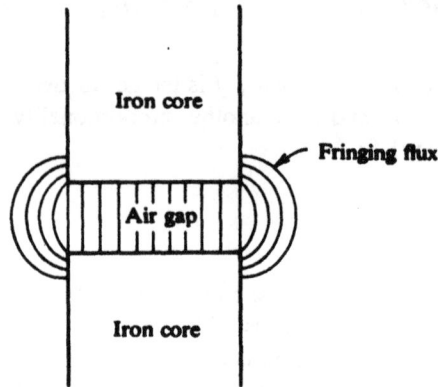

Fig. 1-9

$$W_m = \frac{1}{2} \int_v \mathbf{B} \cdot \mathbf{H} \, dv = \frac{1}{2} \mu \int_v H^2 \, dv = \frac{1}{2\mu} \int_v B^2 \, dv \qquad (1.16)$$

1.8 INDUCTANCE CALCULATIONS

Inductance is defined as flux linkage per unit current;

$$L \equiv \frac{\lambda}{i} = \frac{N\phi}{i} \qquad (1.17)$$

The unit of inductance is the *henry* (H). From (1.17) it is seen that 1 H = 1 Wb/A.

For a magnetic toroid wound with n distinct coils, as shown in Fig. 1-10, n^2 inductances may be defined:

$$L_{pq} \equiv \frac{\text{flux linking the } p\text{th coil due to the current in the } q\text{th coil}}{\text{current in the } q\text{th coil}} = \frac{N_p(k_{pq}\phi_q)}{i_q} \qquad (1.18)$$

where k_{pq}, the fraction of the flux due to coil q that links coil p, is called the *coupling coefficient* between the two coils. By definition, $k_{pq} \leq 1$; a value less than 1 is attributable to leakage flux between the locations of coil p and coil q. When the two subscripts in (1.18) are equal, the inductance is termed *self-inductance*; when unequal, the inductance is termed *mutual inductance* between coils p and q. Inductances are symmetrical; that is, for all p and q,

$$k_{qp} = k_{pq} \quad \text{and} \quad L_{qp} = L_{pq} \qquad (1.19)$$

To express L_{pq} in terms of the magnetic-circuit parameters, we substitute $\phi_q = N_q i_q / \mathfrak{R}$ in (1.18), obtaining

$$L_{pq} = \frac{k_{pq} N_p N_q}{\mathfrak{R}} = k_{pq} N_p N_q \mathcal{P} \qquad (1.20)$$

where $\mathfrak{R}$ is the reluctance of the magnetic circuit and $\mathcal{P}$ is its permeance. We may replace $\mathfrak{R}$ in (1.20) by $l/\mu A$ (for a circuit for which l and A can be defined) to get

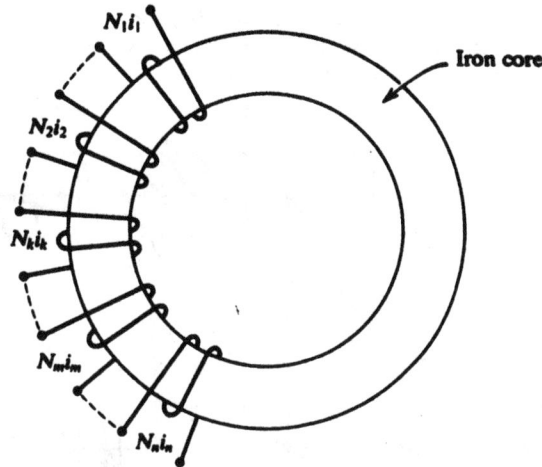

Fig. 1-10

$$L_{pq} = \left(\frac{\mu A}{l}\right) k_{pq} N_p N_q \tag{1.21}$$

Equations (*1.17*) through (*1.21*) can be used for inductance calculations. Alternatively, we may express the energy stored in an inductance L, carrying a current i, as

$$W_m = \frac{1}{2} L i^2 \tag{1.22}$$

and then obtain L by equating the right side of (*1.22*) to the right side of (*1.16*). For an n-coil system, the general relationship is

$$\frac{1}{2}\sum_{p=1}^{n}\sum_{q=1}^{n} L_{pq} i_p i_q = \frac{1}{2}\int_v \mathbf{B}\cdot\mathbf{H}\,dv \tag{1.23}$$

1.9 MAGNETIC CIRCUITS WITH PERMANENT MAGNETS

In Section 1.1 we mentioned that a permanent magnet is the source of a magnetic field. In a magnetic circuit excited by a permanent magnet, the operating conditions of the magnet are largely determined by its position in the circuit. The second-quadrant B-H characteristics (demagnetization curves) of a number of Alnico permanent magnets are shown in Fig. 1-11; Fig. 1-12 shows the characteristics of several ferrite magnets. Commercially available characteristics are still expressed in CGS units (which, if desired, may be converted to SI units by use of Appendix A). Through a point (H_d, B_d) of a demagnetization curve there pass a hyperbola giving the value of the *energy product*, $B_d H_d$, and a ray from the origin (of which only the distal end is shown) giving the value of the *permeance ratio*, B_d/H_d. The significance of the energy product is apparent from (*1.16*), and a permanent magnet is used most efficiently when the energy product is maximized.

Example 1.1 The *remanence*, B_r, of a permanent magnet is the value of B at zero H after saturation; the *coercivity*, H_c, is the value of H to reduce B to zero after saturation. Using Fig. 1-11, find B_r, H_c, and the maximum energy product, $(BH)_{max}$, for Alnico V. Compare with the value listed in Appendix C, Table C-1.

We read B_r and H_c from the vertical and horizontal intercepts, respectively, of the demagnetization curve.

$$B_r = 12.4 \times 10^3 \text{ gauss} = 1.24 \text{ T}$$

$$H_c = 630 \text{ oersteds} = 50 \text{ kA/m}$$

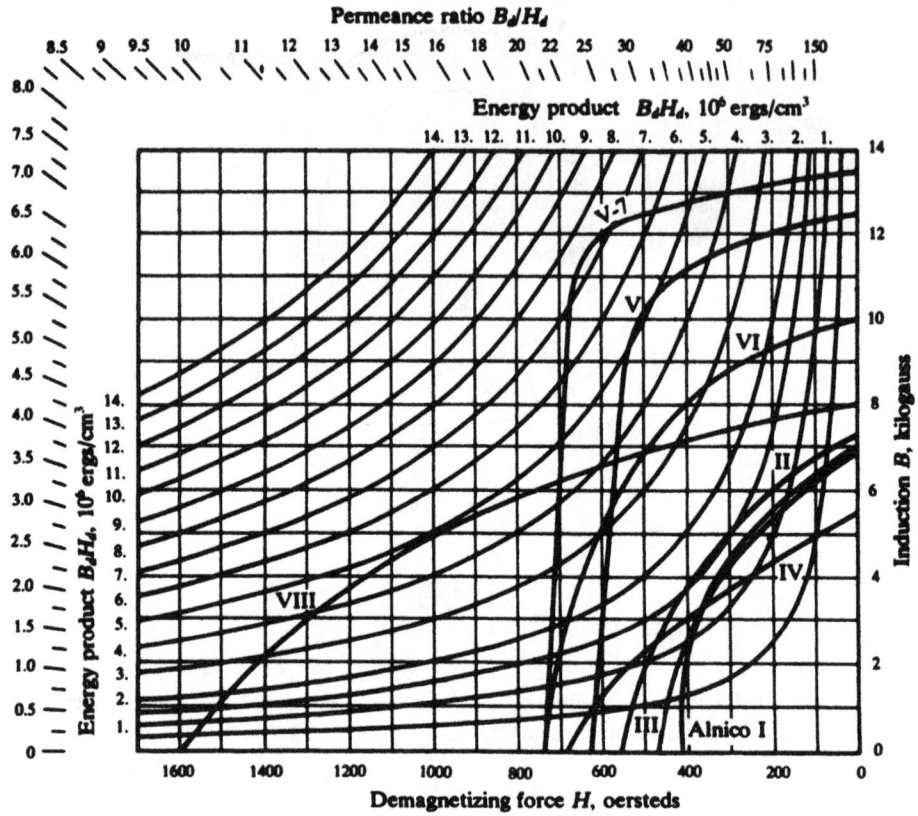

Fig. 1-11. Demagnetization and energy-product curves for Alnico magnets.

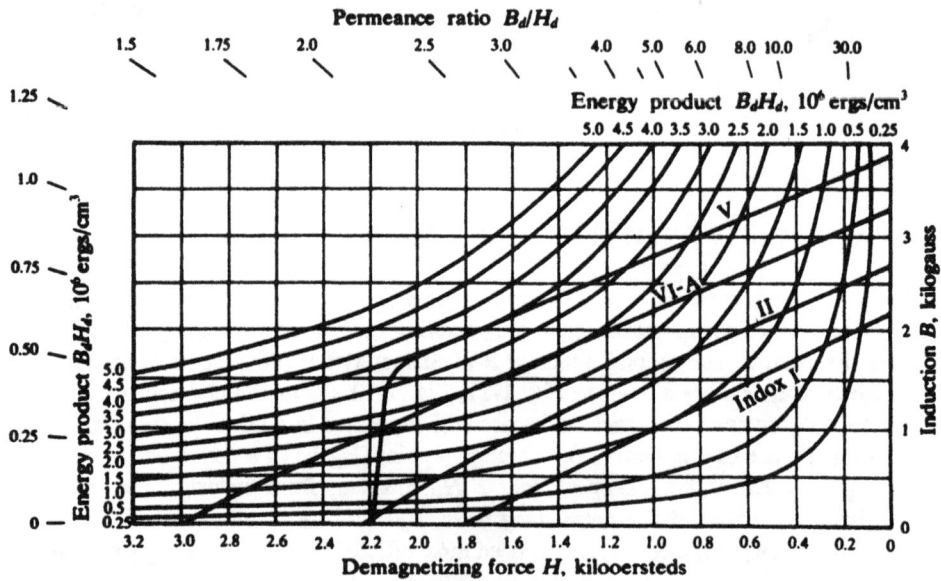

Fig. 1-12. Demagnetization and energy-product curves for Indox ceramic magnets.

where we use Appendix A to convert to SI units. These values are consistent with the ranges given for Alnico V in Table C-1.

The maximum energy product is read from the hyperbola that is just tangent at the knee of the demagnetization curve:

$$(BH)_{max} = 5.2 \times 10^6 \text{ erg/cm}^3 = 520 \text{ kJ/m}^3$$

This value is roughly 10 times larger than the value inferred from Table C-1. However, we must remember that, in the CGS system, (*1.16*) is replaced by

$$W_m = \frac{1}{8\pi} \int_v \mathbf{B} \cdot \mathbf{H} \, dv$$

Thus, CGS energy products are expected to be 4π times as great as SI energy products. Making the adjustment:

$$\frac{1}{4\pi} (BH)_{max} = \frac{5.2 \times 10^6}{4\pi} \text{ ergs/cm}^3 = 41 \text{ kJ/m}^3$$

and now the agreement with Table C-1 is very good.

Once the type of permanent magnet has been chosen, the design approach is as follows. From Ampere's law, for a circuit consisting of an airgap, a portion of a permanent magnet, and another ferromagnetic portion,

$$H_d l_m = H_g l_g + V_{mi} \qquad (1.24)$$

where H_d = magnetic field intensity of the magnet, oersteds
         l_m = length of magnet, cm
         H_g = field intensity in the gap, oersteds = flux density in gap, gauss
         l_g = length of gap, cm
         V_{mi} = reluctance drop in the other ferromagnetic portion, gilberts

Observe that, because μ_0 is unity in the CGS system, H_g and B_g are numerically equal.

The cross-sectional area of the magnet is found from the flux required in the airgap via the relationship

$$B_d A_m = K B_g A_g \qquad (1.25)$$

where B_d = flux density in the magnet, gauss
         A_m = cross-sectional area of magnet, cm^2
         B_g = flux density in the gap, gauss
         A_g = cross-sectional area of gap, cm^2
         K = dimensionless leakage factor

Formulas for the determination of leakage factors for some common configurations are given in Problem 1.17. Solving for A_m in (*1.25*) and for l_m in (*1.24*) (neglecting V_{mi}), and using $H_g = B_g$, we obtain

$$\text{volume} = A_m l_m = \frac{B_g^2 A_g l_g K}{B_d H_d} \text{ (cm}^3) \qquad (1.26)$$

The permeance ratio, shown on Figs. 1-11 and 1-12, is, in the CGS system, the ratio of the equivalent permeance of the external circuit, $\wp_{ge} \equiv A_g K/l_g$, to the permeance of the space occupied by the permanent magnet, $\wp_m \equiv A_m/l_m$. This can be seen by solving (*1.25*) for B_d and (*1.24*) (neglecting V_{mi}) for H_d, and taking the ratio:

$$\frac{B_d}{H_d} = \frac{A_g l_m K}{A_m l_g} = \frac{\mathcal{P}_{ge}}{\mathcal{P}_m} \quad \text{(CGS units)} \qquad (1.27)$$

Equation (1.27) is deceptively simple in appearance, for the task of obtaining analytical expressions for K—and therefore for $\mathcal{P}_{ge}$—is very difficult. Supposing the permeances known, (1.27) plots as a straight line (the *load line*) in the *B-H* plane, and the intersection of this line with the *B-H* curve gives the operating point of the magnet. See Problem 1.33(c).

Solved Problems

1.1. Find the magnetic field intensity due to an infinitely long, straight conductor carrying a current I amperes, at a point r meters away from the conductor.

From Fig. 1-13 and (1.6),

$$\oint \mathbf{H} \cdot d\mathbf{l} = 2\pi r H_\phi = I \quad \text{or} \quad H_\phi = \frac{I}{2\pi r} \quad \text{(A/m)}$$

From the geometry of the problem, the radial and longitudinal components of **H** are zero.

1.2. The conductor of Problem 1.1 carries 100 A current and is located in air. Determine the flux density at a point 0.05 m away from the conductor.

Since $B = \mu_0 H$, from Problem 1.1 we have

$$B_\phi = \mu_0 H_\phi = \frac{\mu_0 I}{2\pi r} = \frac{4\pi \times 10^{-7} \times 100}{2\pi \times 0.05} = 0.4 \text{ mT}$$

1.3. A rectangular loop is placed in the field of the conductor of Problem 1.1 as shown in Fig. 1-14. What is the total flux linking the loop?

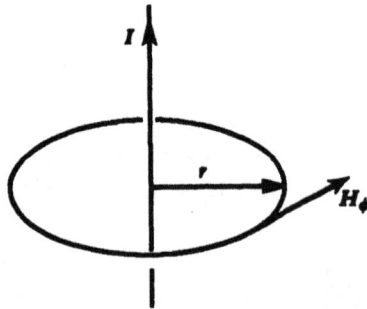

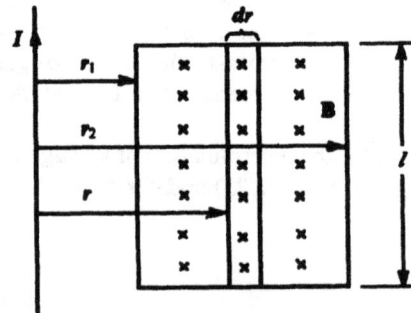

Fig. 1-13   Fig. 1-14

Assuming a medium of permeability μ, we have from Problem 1.1

$$B_\phi = \mu H_\phi = \frac{\mu I}{2\pi r} \quad \text{(T)}$$

The flux $d\phi$ in the elementary area $dA = l\, dr$ is given by

$$d\phi = B_\phi\, dA = \frac{\mu Il}{2\pi}\frac{dr}{r}$$

and

$$\phi = \frac{\mu Il}{2\pi}\int_{r_1}^{r_2}\frac{dr}{r} = \frac{\mu Il}{2\pi}\ln\frac{r_2}{r_1}\quad\text{(Wb)}$$

1.4. A cast steel ring has a circular cross-section 3 cm in diameter and a mean circumference of 80 cm. The ring is uniformly wound with a coil of 600 turns. (a) Estimate the current in the coil required to produce a flux of 0.5 mWb in the ring. (b) If a saw cut creates a 2-mm airgap in the ring, find approximately the airgap flux produced by the current obtained in (a). Find the current which will produce the same flux in the airgap as in (a). Neglect fringing and leakage. Refer to Fig. 1-5(b) for the magnetization characteristic of cast steel.

(a) Ring cross section, $A = \dfrac{\pi}{4}\times 3^2\times 10^{-4} = 7.07\times 10^{-4}\ \text{m}^2$

Core flux density, $B = \dfrac{\phi}{A} = \dfrac{0.5\times 10^{-3}}{7.07\times 10^{-4}} = 0.707\ \text{T}$

From Fig. 1.5(b), @ $B = 0.707$ T, $H = 675$ At/m

Mmf $= \mathscr{F} = Hl = 675\times 0.8 = 540$ At $= NI = 600\ I$

  Thus, $I = \dfrac{540}{600} = 0.9$ A

(b) $\mathscr{R}_{total} = \mathscr{R}_{core} + \mathscr{R}_{air}$. From (a),

$$\mathscr{R}_{core} = \frac{\mathscr{F}}{\phi} = \frac{540}{0.5\times 10^{-3}} = 1.08\times 10^6\ \text{H}^{-1}$$

$$\mathscr{R}_{air} = \frac{q}{\mu_0 A} = \frac{2.0\times 10^{-3}}{4\pi\times 10^{-7}\times 7.07\times 10^{-4}} = 2.25\times 10^6\ \text{H}^{-1}$$

$$\mathscr{R}_{total} = (1.08 + 2.25)10^6 = 3.33\times 10^6\ \text{H}^{-1}$$

Airgap flux, $\phi_{air} = \dfrac{\mathscr{F}}{\mathscr{R}_{total}} = \dfrac{540}{3.33\times 10^6} = 0.162$ mWb

(c) To maintain $\phi = 0.5$ mWb, $\mathscr{F} = R\phi = 3.33\times 10^6\times 0.5\times 10^{-3}$ At

Or $\mathscr{F} = NI = 1665$ At. Thus, $I = \dfrac{1665}{600} = 2.775$ A.

1.5. For the magnetic circuit shown in Fig. 1-15, $N = 10$ turns, $l_g = 0.1$ mm, $l_m = 100$ mm, stacking factor $= 0.9$; the core material is M-19. Calculate I required to establish a 1-T flux density in the airgap. Neglect fringing and leakage.

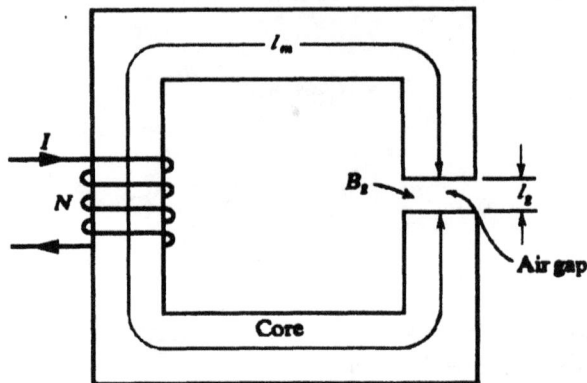

Fig. 1-15

For the airgap:

$$H_g = \frac{B_g}{\mu_0} = \frac{1.0}{4\pi \times 10^{-7}} = 7.95 \times 10^5 \text{ A/m}$$

$$\mathscr{F}_g = H_g l_g = (7.95 \times 10^5)(10^{-4}) = 79.5 \text{ At}$$

For the core:

$$B_m = \frac{B_g}{\text{stacking factor}} = \frac{1}{0.9} = 1.11 \text{ T}$$

and, from Appendix C, Fig. C-1, at 1.11 T we have

$$H_m = 130 \text{ A/m} \quad \text{and} \quad \mathscr{F} = (130)(0.0100) = 13 \text{ At}$$

Then the total required mmf is

$$\mathscr{F}_g + \mathscr{F}_m = 79.5 + 13 = 92.5 \text{ At}$$

from which

$$I = \frac{92.5}{10} = 9.25 \text{ A}$$

1.6. From Appendix C, determine the relative amplitude permeability for (*a*) AISI 1020 and (*b*) M-19, at a flux density of 1 T. ($\mu_0 = 4\pi \times 10^{-7}$ H/m.)

(*a*)

$$\mu_a = \frac{1}{\mu_0}\left(\frac{1}{1600}\right) \approx 500$$

(*b*)

$$\mu_a = \frac{1}{\mu_0}\left(\frac{1}{90}\right) \approx 8800$$

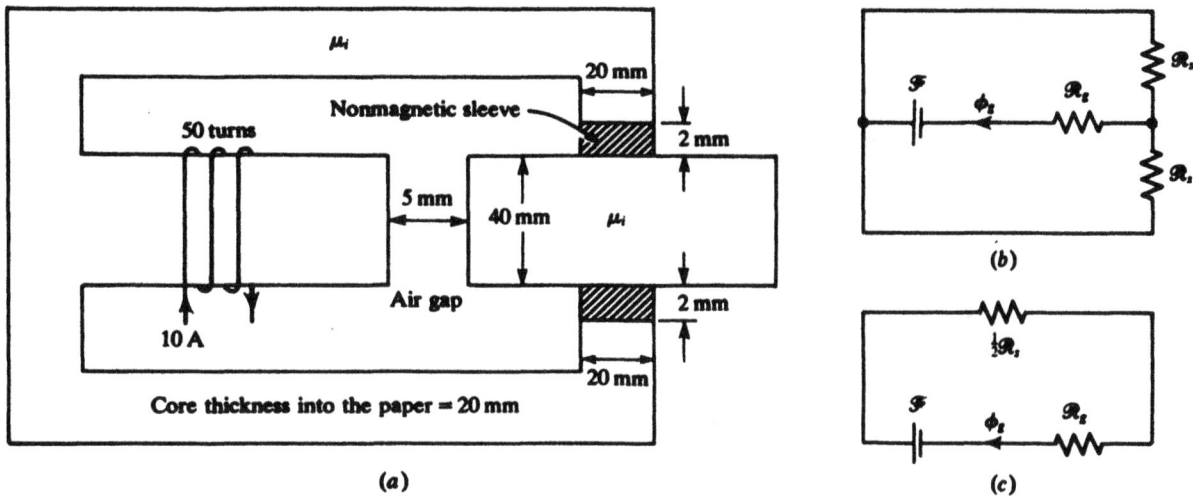

Fig. 1-16

1.7. Assuming an ideal core ($\mu_i \to \infty$), calculate the flux density in the airgap of magnetic circuit shown in Fig. 1-16(a).

> The electrical analog, Fig. 1-16(b), may be reduced to the simpler Fig. 1-16(c). From the latter, and Table 1-1,

$$\mathfrak{R}_g \equiv \text{airgap reluctance} = \frac{5 \times 10^{-3}}{\mu_0 (20 \times 40 \times 10^{-6})} = \frac{50}{8\mu_0}$$

$$\mathfrak{R}_s \equiv \text{sleeve reluctance} = \frac{2 \times 10^{-3}}{\mu_0 (20 \times 20 \times 10^{-6})} = \frac{20}{4\mu_0}$$

$$\mathfrak{R}_t \equiv \text{total reluctance} = \mathfrak{R}_g + \frac{1}{2}\mathfrak{R}_s = \frac{70}{8\mu_0}$$

$$\phi_g \equiv \text{airgap flux} = \frac{\mathfrak{F}}{\mathfrak{R}_t} = \frac{(50)(10)}{70/8\mu_0} = \frac{400\mu_0}{7}$$

$$B_g \equiv \text{airgap flux density} = \frac{\phi_g}{A_g} = \frac{400\mu_0/7}{20 \times 40 \times 10^{-6}}$$

Substituting $\mu_0 = 4\pi \times 10^{-7}$, we get $B_g = 90$ mT, or 900 gauss.

1.8. A composite magnetic circuit of varying cross section is shown in Fig. 1-17(a); the iron portion has the B-H characteristic of Fig. 1-17(b). Given: $N = 100$ turns; $l_1 = 4l_2 = 40$ cm; $A_1 = 2A_2 = 10$ cm^2; $l_g = 2$ mm; leakage flux, $\phi_l = 0.01$ mWb. Calculate I required to establish an air-gap flux density of 0.6 T.

> Corresponding to $B_g = 0.6$ T,

$$H_g = \frac{0.6}{\mu_0} = 4.78 \times 10^5 \text{ A/m}$$

$$\mathfrak{F}_g = (4.78 \times 10^5)(2 \times 10^{-3}) = 956 \text{ At}$$

$$B_{ll} = B_g = 0.6 \text{ T}$$

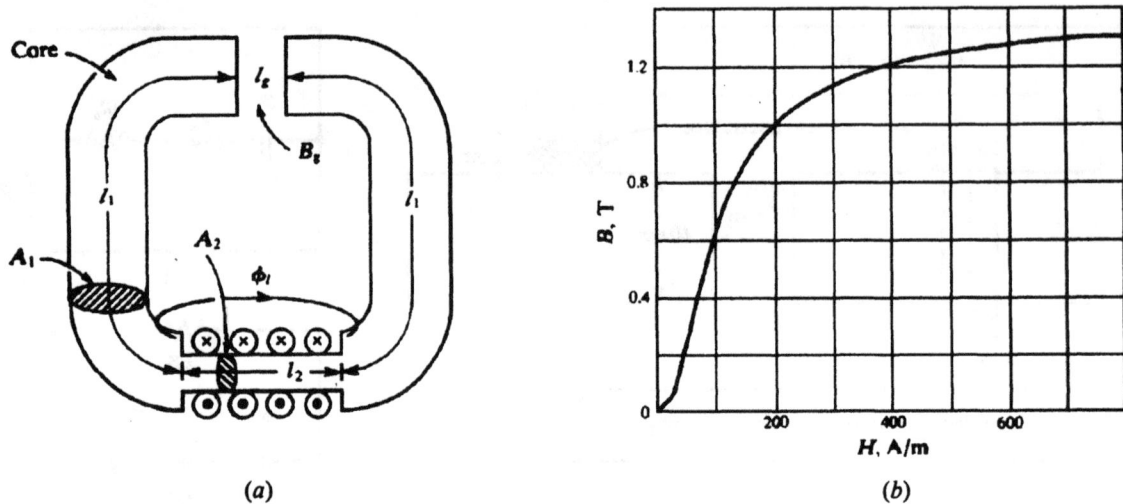

(a)   (b)

Fig. 1-17

From Fig. 1-17(b) at $B = 0.6$ T, $H = 100$ A/m. Thus, for the two lengths l_1 we have

$$\mathscr{F}_{l1} = (100)(0.40 + 0.40) = 80 \text{ At}$$

The flux in the airgap, ϕ_g, is given by

$$\phi_g = B_g A_1 = (0.6)(10 \times 10^{-4}) = 0.6 \text{ mWb}$$

The total flux produced by the coil, ϕ_c, is the sum of the airgap flux and the leakage flux:

$$\phi_c = \phi_g + \phi_l = 0.6 + 0.01 = 0.61 \text{ mWb}$$

The flux density in the portion l_2 is, therefore,

$$B_2 = \frac{\phi_c}{A_2} = \frac{0.61 \times 10^{-3}}{5 \times 10^{-4}} = 1.22 \text{ T}$$

For this flux density, from Fig. 1-17(b), $H = 410$ A/m and

$$\mathscr{F}_{l2} = (410)(0.10) = 41 \text{ At}$$

The total required mmf, $\mathscr{F}$, is thus

$$\mathscr{F} = \mathscr{F}_g + \mathscr{F}_{l1} + \mathscr{F}_{l2} = 956 + 80 + 41 = 1077 \text{ At}$$

For $N = 100$ turns, the desired current is, finally,

$$I = \frac{1077}{100} = 10.77 \text{ A}$$

1.9.   Draw an electrical analog for the magnetic circuit shown in Fig. 1-17(a).

        See Fig. 1-18.

1.10.   Calculate the (total) self-inductance and the leakage inductance of the coil shown in Fig. 1-17(a).

        From Problem 1.8 the total flux produced by the coil is $\phi_c = 0.61$ mWb and $I = 10.77$ A. Hence

$$L = \frac{N\phi_c}{I} = \frac{(100)(0.61 \times 10^{-3})}{10.77} = 5.66 \text{ mH}$$

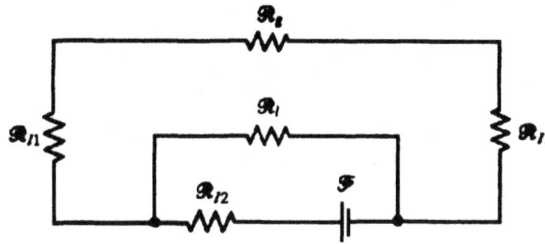

Fig. 1-18

and
$$L_l = \frac{N\phi_i}{I} = \frac{(100)(0.01 \times 10^{-3})}{10.77} = 0.093 \text{ mH}$$

1.11. Determine the magnetic energy stored in the iron and in the airgap of the magnetic circuit of Fig. 1-17(*a*).

From (*1.16*),

$$W_{air} = \frac{1}{2\mu_0} B_g^2 \times vol_{gap} = \frac{(0.6)^2}{2\mu_0} [(10 \times 10^{-4})(2 \times 10^{-3})] = 0.286 \text{ J}$$

From (*1.22*) and Problem 1.10,

$$W_{iron} = \frac{1}{2} LI^2 - W_{air} = \frac{1}{2} N\phi_c I - W_{air} = 0.328 - 0.286 = 0.042 \text{ J}$$

1.12. If the stacking factor is 0.8 and B_g remains 0.6 T, determine the flux densities in the various portions of the magnetic circuit of Fig. 1-17(*a*).

$$B_g = 0.6 \text{ T}$$

$$B_{l1} = \frac{B_g}{\text{stacking factor}} = \frac{0.6}{0.8} = 0.75 \text{ T}$$

$$B_{l2} = \frac{B_2}{\text{stacking factor}} = \frac{1.22}{0.8} = 1.525 \text{ T}$$

where the value of B_2 is taken from Problem 1.8.

1.13. A toroid of rectangular cross section is shown in Fig. 1-19. The mean diameter is large compared to the core thickness in the radial direction, so that the core flux density is uniform. Derive an expression for the inductance of the toroid, and evaluate it if $r_1 = 80$ mm, $r_2 = 100$ mm, $a = 20$ mm, and $N = 200$ turns. The core relative permeability is 900.

flux linkage: $\lambda = N\phi$

flux: $\phi = \frac{Ni}{\mathfrak{R}} = \frac{\mu A Ni}{2\pi r} = \frac{\mu a(r_2 - r_1)Ni}{\pi(r_2 + r_1)}$

since $A = a(r_2 - r_1)$ and $r = (r_2 + r_1)/2$. Hence

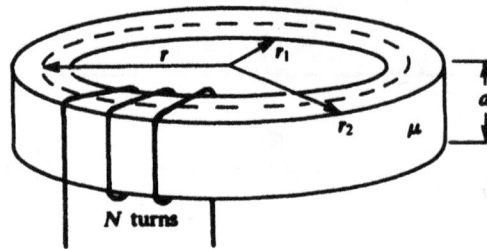

Fig. 1-19

$$L = \frac{\lambda}{i} = \frac{\mu a(r_2 - r_1)N^2}{\pi(r_2 + r_1)}$$

Substituting the numerical values, we obtain

$$L = \frac{(900\mu_0)(20 \times 10^{-3})(20 \times 10^{-3})(200)^2}{\pi(180 \times 10^{-3})} = 32 \text{ mH}$$

1.14. The density of the core material of the toroid of Fig. 1-19 is 7.88×10^3 kg/m³ and the core is wound with a round wire, AWG size 8. What is the total mass of the toroid?

volume of core	=	$(\pi r_2^2 - \pi r_1^2)a = \pi[(0.100)^2 - (0.080)^2](0.020) = 72\pi \times 10^{-6}$ m³
mass of core	=	$(7.88 \times 10^3)(72\pi \times 10^{-6}) = 1.782$ kg
mean length per turn	=	$2(a + r_2 - r_1) + (10\%$ for bending around corners$) = 0.088$ m
total length of coils	=	$(200)(0.088) = 17.6$ m

From Appendix B, No. 8 wire weighs 50.2 lb/1000 ft, or 0.0747 kg/m. Thus

mass of coil	$= (17.6)(0.0747) = 1.315$ kg
total mass of toroid	$= 1.782 + 1.315 = 3.097$ kg

1.15. For the toroid shown in Fig. 1-19, (a) derive an expression for the magnetic field intensity, $H(r)$. (b) What is the core flux if $\mu_r = 1$? (c) If the core flux density is assumed to be uniform, and equal to its value at the (arithmetic) mean radius, what percent error would be made in the computation of the core flux by this approximation as compared to the calculation in (b)? (d) If the geometric mean radius is used instead of the arithmetic mean, what is the percent error?

(a)
$$H = \frac{NI}{2\pi r}$$

(b)
$$\phi = \int_{r_1}^{r_2} \mu_0 \frac{NI}{2\pi r} a \, dr = \mu_0 \frac{aNI}{2\pi} \ln \frac{r_2}{r_1}$$

(c)   Using the arithmetic mean, $B \approx \mu_0 NI/\pi(r_2 + r_1)$, and so

$$\phi \approx \mu_0 \frac{aNI}{2\pi} \frac{2(r_2 - r_1)}{r_2 + r_1}$$

Let $r_2/r_1 \equiv b$; then

$$\text{percent error} = 100 \left[1 - \frac{2(b - 1)}{(b + 1) \ln b} \right]$$

For example, if $b = 2$, percent error $= 3.9\%$.

(d)   Using the geometric mean, $B \approx \mu_0 NI/2\pi \sqrt{r_2 r_1}$, and so

$$\phi \approx \mu_0 \frac{aNI}{2\pi} \frac{r_2 - r_1}{\sqrt{r_2 r_1}}$$

$$\text{percent error} = 100 \left[\frac{b - 1}{\sqrt{b} \ln b} - 1 \right]$$

If $b = 2$, percent error $= 2\%$.

1.16.   Sometimes the *B-H* curve of a core material can be expressed by the *Froelich equation*,

$$B = \frac{aH}{b + H} \tag{1.28}$$

where a and b are constants of the material. Let $a = 1.5$ T and $b = 100$ A/m. A magnetic circuit consists of two parts (in series), of lengths l_1 and l_2 and cross-sectional areas A_1 and A_2. If $A_1 = 25$ cm$^2 = 2A_2$ and $l_1 = 25$ cm $= \frac{1}{2}l_2$, and if the core carries an mmf of 1000 At, calculate the core flux.

From (*1.28*),

$$B = \frac{1.5 H}{100 + H} \quad \text{(T)}$$

and for the magnetic circuit,

$$\mathcal{F} = H_1 l_1 + H_2 l_2 = \frac{1}{4} H_1 + \frac{1}{2} H_2 = 1000$$

and

$$\phi = B_1 A_1 = B_2 A_2 \quad \text{or} \quad 2B_1 = B_2 \quad \text{or} \quad \frac{3.0 H_1}{100 + H_1} = \frac{1.5 H_2}{100 + H_2}$$

Eliminating H_1 between the above two equations yields

$$-3H_2^2 + 5250 + 1200000 = 0 \quad \text{or} \quad H_2 = 1954.6 \text{ A/m}$$

Thus

$$B_2 = \frac{(1.5)(1954.6)}{100 + 1954.6} = 1.427 \text{ T}$$

and
$$\phi = (1.427)\left(\frac{25}{2} \times 10^{-4}\right) = 1.784 \text{ mWb}$$

     If the *B-H* characteristic had been given in graphical form, this problem could only be solved by trial and error. See Problem 1.31.

1.17. Determine the length, *b*, and cross-sectional area, A_m, of the magnet in Fig. 1-20(*a*) to produce a flux density of 2500 gauss in the airgap. The permanent magnet to be used is Alnico V; the dimensions of Fig. 1-20(*a*) are as follows: $l_g = 0.4$ cm, $c = 6.0$ cm, gap area = 4.0 cm^2 (2 cm × 2 cm). We assume that the reluctance in the soft-iron parts of the circuit is negligible, giving a reluctance drop, V_{mi}, of zero; we estimate that the leakage factor is 4.0 and that the magnet is to be operated at maximum energy product (knee of the demagnetization curve in Fig. 1-11).

    From (*1.25*),

$$A_m = \frac{B_g A_g K}{B_d} = \frac{(2500)(4)(4.0)}{10.5 \times 10^3} = 3.8 \text{ cm}^2$$

From (*1.24*), noting that $H_g = B_g$ in CGS units,

$$b = l_m = \frac{(2500)(0.4)}{450} = 2.22 \text{ cm}$$

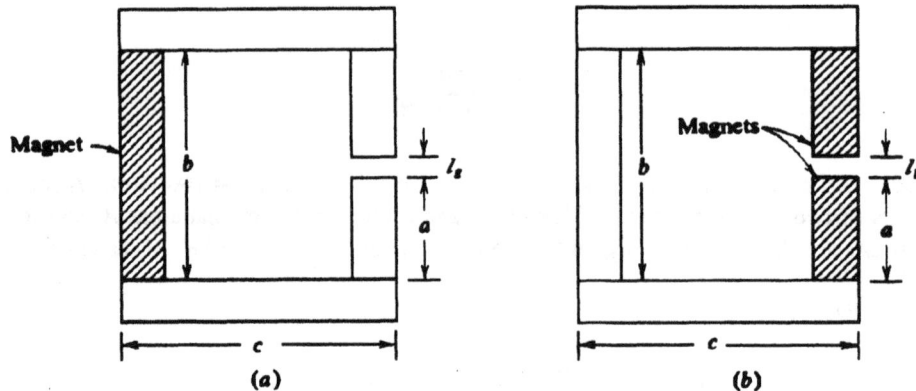

Fig. 1-20

    We must now check our estimate of the leakage factor. A leakage factor for the configuration of Fig. 1-20(*a*) is given by

$$K = 1 + \frac{l_g}{A_g}\left[1.7 C_a\left(\frac{a}{a + l_g}\right) + 1.4 c\sqrt{\frac{C_c}{b}} + 0.67 C_b\right] \qquad (1.29)$$

where C_a, C_b, and C_c are the cross-sectional perimeters of the circuit portions whose lengths are *a*, *b*, and *c*, respectively. The factor 0.67 in (*1.29*) arises from the fact that permanent magnets have a "neutral zone" that does not contribute to leakage. Substituting $b = 2.2$ cm, $a = (b - l_g)/2 = 0.91$ cm, $c = 6.0$ cm, $C_a = (4)(2) = 8$ cm, $C_c = 8.0$ cm, and $C_b = 4\sqrt{3.8} = 7.80$ cm into (*1.29*) gives

$$K = 4.062$$

This value could now be put back into (1.25), giving a slightly different value of A_m. This, in turn would change C_b in (1.29), resulting in a new value of leakage factor. A few iterations of these formulas are usually necessary to obtain a consistent set of dimensions for the total magnetic circuit.

The high value for leakage factor obtained for the configuration of Fig. 1-20(a) indicates that this is not a very efficient magnetic circuit. A much more efficient use of the permanent magnet is to locate it adjacent to the airgap, as shown in Fig. 1-20(b). The leakage factor for Fig. 1-20(b) is

$$K = 1 + \frac{l_g}{A_g} 0.67 C_a \left[1.7 \left(\frac{0.67 C_a}{0.67a + l_g} \right) + \frac{l_g}{2a} \right] \qquad (1.30)$$

Using the same dimensions for all sections of the circuit in Fig. 1-20(b) as were used in Fig. 1-20(a) (even though this might result in an oversized permanent magnet), we obtain from (1.30)

$$K = 1.624$$

1.18. An inductor, made of a highly permeable material, has N turns. The dimensions of the core and the coil are as shown in Fig. 1-21. Calculate the input power to the coil to establish a given flux density B in the airgap. The winding space-factor of the coil is k_s and its conductivity is σ.

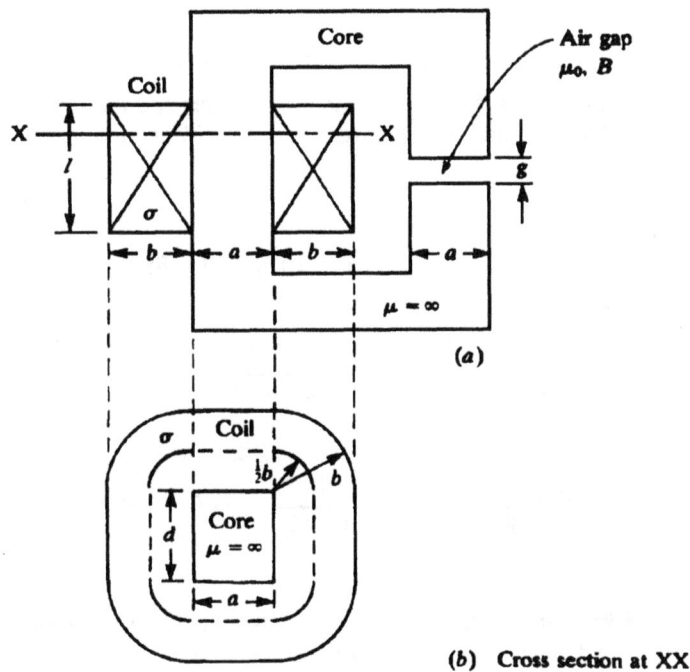

Fig. 1-21

From Fig. 1-21(b), the mean length per turn is

$$l = 2a + 2d + 4 \left(\frac{1}{4} \right) \left(2\pi \frac{b}{2} \right) = 2 \left(a + d + \pi \frac{b}{2} \right)$$

and the total length of wire making the coil is lN. Let A_c denote the cross-sectional area of the wire; then its resistance is

$$R = \frac{lN}{\sigma A_c}$$

giving an input power of

$$P_i = I^2 R = \frac{I^2 l N}{\sigma A_c}$$

But

$$\mathscr{F} = NI = \phi \mathscr{R} = BA \frac{g}{\mu_0 A} = \frac{Bg}{\mu_0} \quad \text{or} \quad I = \frac{B_g}{\mu_0 N}$$

Substituting above yields

$$P_i = \frac{B^2 g^2 l}{\mu_0^2 \sigma N A_c}$$

The total volume of the wire is $(bcl)k_s = lNA_c$. Hence, finally,

$$P_i = \frac{B^2 g^2 l}{\mu_0^2 \sigma bck_g}$$

1.19. The inductor of Problem 1.18 is made of magnet wire. If the dimensions in Fig. 1-21 are

$$a = b = c = d = 25 \text{ mm} \qquad g = 2 \text{ mm}$$

and the core flux density is 0.8 T, calculate the input power and the number of turns. Assume $k_s = 0.8$, $\sigma = 5.78 \times 10^7$ S/m, and a coil current of 1 A. (*Note:* 1 S = 1 Ω^{-1}.)

From Problem 1.18,

$$l = 2\left(1 + 1 + \frac{\pi}{2}\right)(25 \times 10^{-3}) = 0.1785 \text{ m}$$

$$P_i = \frac{(0.8)^2 (2 \times 10^{-3})^2 (0.1785)}{(4\pi \times 10^{-7})^2 (5.78 \times 10^7)(25 \times 10^{-3})^2 (0.8)} = 10 \text{ W}$$

Also from Problem 1.18

$$N = \frac{Bg}{\mu_0 I} = \frac{(0.8)(2 \times 10^{-3})}{(4\pi \times 10^{-7})(1)} = 1273 \text{ turns}$$

1.20. For the inductor of Problem 1.19, find the area of the conductor cross section. What is the time constant of the coil and at what voltage may it be operated?

From $P_i = I^2 R = 10$ W and $I = 1$ A, $R = 10 \Omega$. Then the operating voltage is

$$V = IR = (1)(10) = 10 \text{ V}$$

Also, from $R = lN/\sigma A_c$,

$$A_c = \frac{lN}{\sigma R} = \frac{(0.1785)(1273)}{(5.78 \times 10^7)(10)} = 0.393 \times 10^{-6} \text{ m}^2 = 0.393 \text{ mm}^2$$

The flux linking the coil is $\phi = Bad$, and so

$$L = \frac{Bad}{I} = \frac{(0.8)(25 \times 10^{-3})^2}{1} = 0.5 \text{ mH}$$

and the time constant is

$$\tau = \frac{L}{R} = \frac{0.5 \times 10^{-3}}{10} = 50 \ \mu s$$

Notice that from the value of A_c a suitable wire size can be determined.

Supplementary Problems

1.21. From Appendix C, Fig. C-1, determine the relative amplitude permeability at a flux density of 1.2 T for (a) M-19 and (b) 48 NI. *Ans.* (a) 5457; (b) 9550

1.22. Replot the *B-H* curve for M-19 on rectangular coordinate paper. (Figure C-1 of Appendix C is plotted on semilog paper.) Identify the three ranges of permeability, I, II, and III, of Fig. 1-5. *Ans.* For range II: $0.4 \le B \le 0.8$ T

1.23. The magnetic circuit of Fig. 1-22 has the *B-H* characteristic of Fig. 1-17(b). Calculate the mmf of the coil to establish a 1-T flux density in the airgap. *Ans.* 902 At

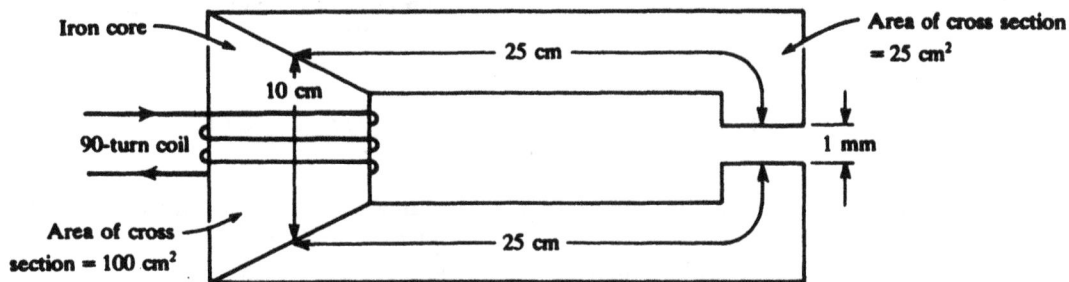

Fig. 1-22

1.24. The coil of Fig. 1-22 has 90 turns. For the data of Problem 1.23, determine (a) the energy stored in the coil, (b) the energy stored in the airgap, (c) the energy stored in iron. *Ans.* (a) 1.13 J; (b) 0.995 J; (c) 0.135 J

1.25. Calculate the inductance of the coil of Fig. 1-22, (a) excluding the effect of the iron core (i.e., assuming the core to be infinitely permeable) and (b) including the effect of the iron core. *Ans.* (a) 25.45 mH; (b) 22.45 mH

1.26. The toroid of Fig. 1-19 is cut to make an airgap 2 mm in length. Given: $r = 500$ mm; core composed of 0.2-mm strips of 48 NI magnetic material; core flux density is 0.6 T; and $N = 100$ turns. Including the effect of stacking factor, calculate the coil current for the given flux density. *Ans.* 8.66 A

1.27. A system of three coils on an ideal core is shown in Fig. 1-23, where $N_1 = N_3 = 2N_2 = 500$ turns, $g_1 = 2g_2 = 2g_3 = 4$ mm, and $A = 1000$ mm^2. Calculate (a) the self-inductance of coil N_1 and (b) the mutual inductance between coils N_2 and N_3. *Ans.* (a) 62.83 mH; (b) 31.42 mH

1.28.   If gap g_1 (Fig. 1-23) is closed, what are the mutual inductances between (a) N_1 and N_2, (b) N_2 and N_3, and (c) N_3 and N_1?   *Ans.* (a) 78.54 mH; (b) 0; (c) 157.08 mH

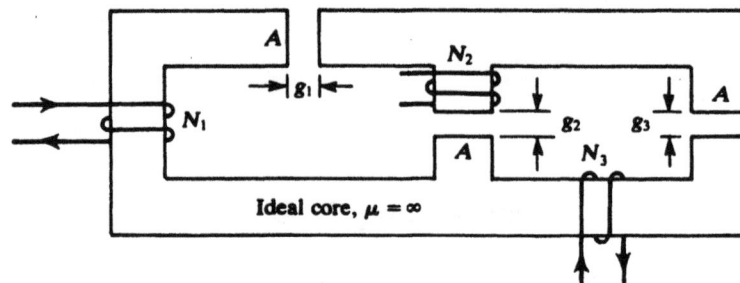

Fig. 1-23

1.29.   The coils of Problem 1.28 are connected in series (with all mutual inductances being positive) and carry a current of 10 A. What is the total energy stored in the entire magnetic circuit?
Ans. 37.3 J

1.30.   The toroid of Fig. 1-19 is made of 0.019-in-thick silicon steel laminations having the characteristics shown in Appendix C, Fig. C-2. The density of the material is 7.88×10^3 kg/m^3 and the dimensions of the core are $r_1 = 100$ mm, $r_2 = 120$ mm, and $a = 40$ mm. If the maximum core flux density is 1 T at 150 Hz, determine the total core loss.   *Ans.* 19.2 W

1.31.   Solve Problem 1.16 using Fig. 1-17(b) instead of (1.28).   *Ans.* $\phi \approx 1.8$ mWb

1.32.   A toroid has a core of square cross section, 2500 mm^2 in area, and a mean diameter of 250 mm. The core material is of relative permeability 1000. (a) Calculate the number of turns to be wound on the core to obtain an inductance of 1 H. (b) If the coil thus wound carries 1 A, what are the values of H and B at the mean radius of the core?   *Ans.* (a) 500 turns; (b) 636 A/m, 0.8 T

1.33.   In the airgap of a C-shaped permanent magnet, made of Alnico V (Fig. 1-11), it is desired to have a 5000-gauss flux density. The length of the airgap is 2 cm and its cross-sectional area is 4 cm$_2$. (a) Calculate the minimum length of the magnet (while operating at maximum energy product). (b) Assuming a leakage factor of 10, determine the area of cross section of the magnet. (c) Suppose that the airgap flux density is unknown, but the results of (a) and (b) still hold. Find the operating flux density of the magnet by the load line method.
Ans. (a) 18.7 cm; (b) 19 cm^2; (c) 10.35 kilogauss

1.34.   A toroid is constructed of 48 NI alloy. The mean length of the toroid is 250 mm and its cross-sectional area is 200 mm^2. If the toroid is to be used in an application requiring a flux of 0.2 mWb, (a) what mmf must be applied to the toroid? (b) It is desired that the coil have an inductance of 10 mH when the flux is 0.2 mWb. Determine the number of turns in the coil.
Ans. (a) 3.75 At; (b) 14 turns

1.35. The magnetic circuit of Fig. 1-24 is made of transformer plates having the *B-H* characteristic of Fig. 1-5(*b*). The magnetic shunt has a relative permeability of 18. The entire magnetic circuit has a uniform cross-section of 10 cm². Other dimensions are: *ab* = *cd* = 10 cm; *befc* = 20 cm; *bc* = 10 cm; *ad* = airgap = 0.1 cm. Calculate (*a*) the magnetomotive force of the *N*-turn coil to establish 1.0 T flux density in the airgap; and (*b*) the inductance of the coil, if *N* = 1000.

Ans. (*a*) 1096 At; (*b*) 10.95 mH

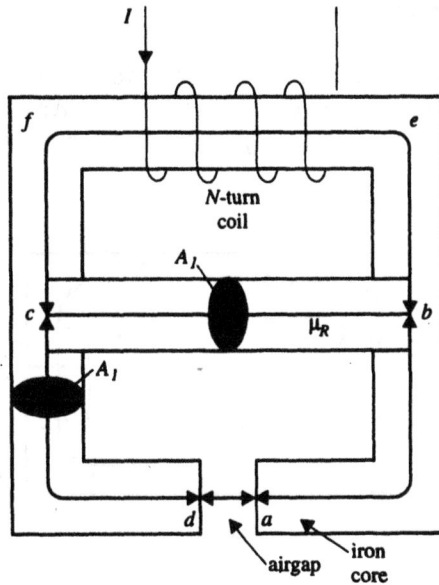

Fig. 1-24

Chapter 2

Power Transformers

2.1 TRANSFORMER OPERATION AND FARADAY'S LAW

A *transformer* is an electromagnetic device having two or more stationary coils coupled through a mutual flux. A two-winding *ideal transformer* is shown in Fig. 2-1. An ideal transformer is assumed to have (i) an infinitely permeable core with no losses, (ii) lossless electrical windings, and (iii) no leakage fluxes.

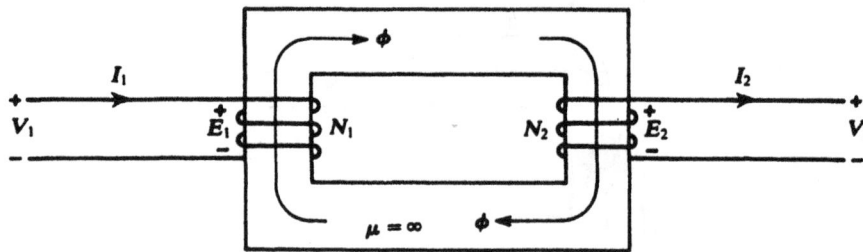

Fig. 2-1

The basic components of a transformer are the core, the *primary winding* N_1, and the *secondary winding* N_2. The action of a transformer is based on *Faraday's law of electromagnetic induction*, according to which a time-varying flux linking a coil induces an emf (voltage) in it. Thus, referring to Fig. 2-1, if ϕ is the flux linking the N_1-turn winding, then its induced voltage, e_1, is given by

$$e_1 = N_1 \frac{d\phi}{dt} \quad \text{(V)} \tag{2.1}$$

The direction of e_1 is such as to produce a current that gives rise to a flux which opposes the flux change $d\phi/dt$ (*Lenz's law*). The transformer being ideal, $e_1 = v_1$; that is, the instantaneous values of the induced voltage and the terminal voltage are equal. Hence, from (*2.1*),

$$\phi = \frac{1}{N_1} \int v_1 \, dt \quad \text{(Wb)} \tag{2.2}$$

Because only the time-variation of ϕ is important, we ignore the constant of integration in (*2.2*).
 If

$$\phi = \phi_m \sin \omega t \tag{2.3}$$

then, from (*2.1*),

$$e_1 = \omega N_1 \phi_m \cos \omega t \tag{2.4}$$

Similarly, the voltage, e_2, induced in the secondary is given by

$$e_2 = \omega N_2 \phi_m \cos \omega t \tag{2.5}$$

From (*2.4*) and (*2.5*),

$$\frac{e_1}{e_2} = \frac{N_1}{N_2}$$

which may also be written in terms of rms values as

$$\frac{E_1}{E_2} = \frac{N_1}{N_2} \equiv a \qquad (2.6)$$

where a is known as the *turns ratio*. In the case that $N_2 > N_1$, one conventionally writes $1/a$ instead of a in (2.6); thus, the turns ratio is always greater than 1.

Because the transformer is ideal, the net mmf around the magnetic circuit must be zero; that is, if I_1 and I_2 are the primary and secondary currents, respectively, the $N_1 I_1 - N_2 I_2 = 0$, or

$$\frac{I_2}{I_1} = \frac{N_1}{N_2} \equiv a \qquad (2.7)$$

From (2.6) and (2.7) it can be shown that if an impedance Z_2 is connected to the secondary, the impedance Z_1 seen at the primary satisfies

$$\frac{Z_1}{Z_2} = \left(\frac{N_1}{N_2}\right)^2 \equiv a^2 \qquad (2.8)$$

2.2 EMF EQUATION OF A TRANSFORMER

For a sinusoidal flux, the rms value of the induced emf in the primary is, from (2.4),

$$E_1 = \frac{\omega N_1 \phi_m}{\sqrt{2}} = 4.44 f N_1 \phi_m \quad (V) \qquad (2.9)$$

where $f = \omega/2\pi$ is the frequency in Hz.

2.3 TRANSFORMER LOSSES

In Section 2.1 we have considered an ideal transformer, which was assumed to have no losses. Obviously, an actual transformer has the following losses:

1. Core losses, which include the hysteresis and eddy-current losses (see Section 1.4).

2. Resistive ($I^2 R$) losses in the primary and secondary windings.

2.4 EQUIVALENT CIRCUITS OF NONIDEAL TRANSFORMERS

A nonideal transformer differs from an ideal transformer in that the former has hysteresis and eddy-current (or core) losses, and has resistive ($I^2 R$) losses in its primary and secondary windings. Furthermore, the core of a nonideal transformer is not perfectly permeable, and the transformer core requires a finite mmf for its magnetization. Also, not all fluxes link with the primary and secondary windings simultaneously because of leakages. Referring to Fig. 2-2, we observe that R_1 and R_2 are the respective resistances of the primary and secondary windings. The flux ϕ_c, which replaces the flux ϕ of Fig. 2-1, is called the *core flux* or *mutual flux*, as it links both the primary and secondary windings. The primary and secondary leakages fluxes are shown as ϕ_{l1} and ϕ_{l2}, respectively. Thus in Fig. 2-2 we have accounted for all the imperfections listed above, except the core losses. We will include the core losses as well as the rest of the imperfections in the equivalent circuit of a nonideal transformer. This circuit is also known as the *exact equivalent circuit*, as it differs from the idealized equivalent circuit and the various approximate equivalent circuits. We now proceed to derive these circuits.

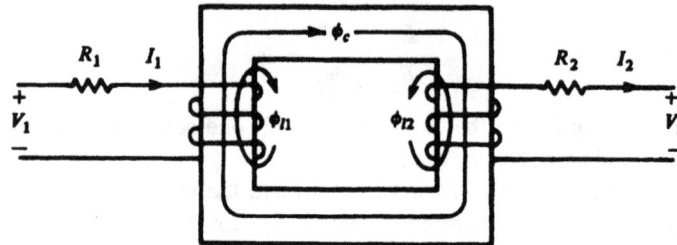

Fig. 2-2

An equivalent circuit of an ideal transformer is shown in Fig. 2-3(a). When the nonideal effects of winding resistances, leakage reactances, magnetizing reactance, and core losses are included, the circuit of Fig. 2-3(a) is modified to that of Fig. 2-3(b), where the primary and the secondary are coupled by an ideal transformer. By use of (2.6), (2.7), and (2.8), the ideal transformer may be removed from Fig. 2-3(b) and the entire equivalent circuit may be referred either to the primary, as shown in Fig. 2-4(a), or to the secondary, as shown in Fig. 2-4(b).

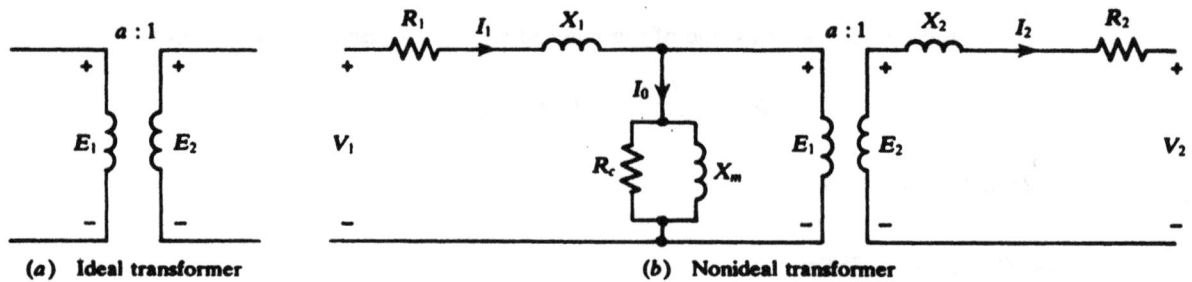

(a) Ideal transformer   (b) Nonideal transformer

Fig. 2-3

A phasor diagram for the circuit Fig. 2-4(a), for lagging power factor, is shown in Fig. 2-5. In Figs. 2-3, 2-4, and 2-5 the various symbols are:

a ≡ turns ratio (>1)
E_1 ≡ primary induced voltage
E_2 ≡ secondary induced voltage
V_1 ≡ primary terminal voltage
V_2 ≡ secondary terminal voltage
I_1 ≡ primary current
I_2 ≡ secondary current
I_0 ≡ no-load (primary) current
R_1 ≡ resistance of the primary winding
R_2 ≡ resistance of the secondary winding
X_1 ≡ primary leakage reactance
X_2 ≡ secondary leakage reactance
I_m, X_m ≡ magnetizing current and reactance
I_c, R_c ≡ current and resistance accounting for the core losses

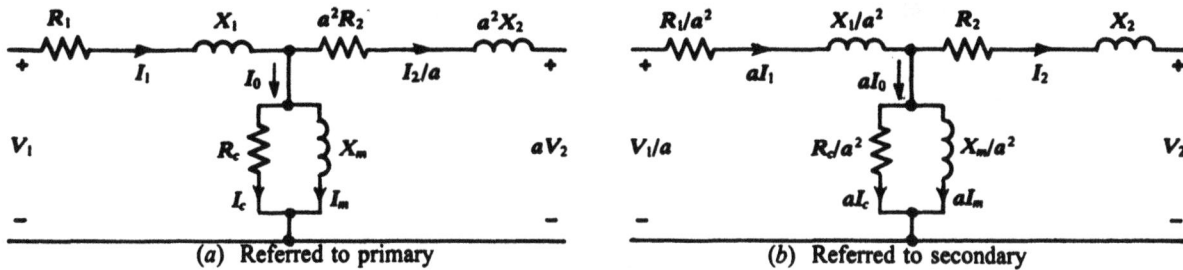

Fig. 2-4. Equivalent circuits of a nonideal transformer.

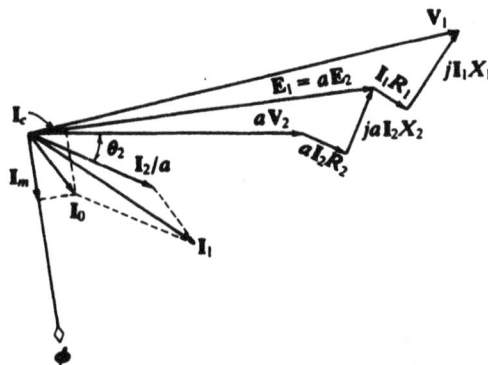

Fig. 2-5. Phasor diagram corresponding to Fig. 2-4(*a*).

2.5 TESTS ON TRANSFORMERS

Transformer performance characteristics can be obtained from the equivalent circuits of Section 2.4. The circuit parameters are determined either from design data or from test data. The two common tests are as follows.

Open-Circuit (or No-Load) Test

Here one winding is open-circuited and voltage—usually, rated voltage at rated frequency—is applied to the other winding. Voltage, current, and power at the terminals of this winding are measured. The open-circuit voltage of the second winding is also measured, and from this measurement a check on the turns ratio can be obtained. It is usually convenient to apply the test voltage to the winding that has a voltage rating equal to that of the available power source. In step-up voltage transformers, this means that the open-circuit voltage of the second winding will be higher than the applied voltage, sometimes much higher. Care must be exercised in guarding the terminals of this winding to ensure safety for test personnel and to prevent these terminals from getting close to other electrical circuits, instrumentation, grounds, and so forth.

In presenting the no-load parameters obtainable from test data, it is assumed that voltage is applied to the primary and the secondary is open-circuited. The no-load power loss is equal to the wattmeter reading in this test; core loss is found by subtracting the ohmic loss in the primary, which is usually small and may be neglected in some cases. Thus, if P_0, I_0, and V_0 are the input power, current, and voltage, then the core loss is given by

$$P_c = P_0 - I_0^2 R_1 \qquad (2.10)$$

The primary induced voltage is given in phasor form by

$$\mathbf{E}_1 = V_0 \angle 0° - (I_0 \angle \theta_0)(R_1 + jX_1) \tag{2.11}$$

where $\theta_0 \equiv$ no-load power-factor angle $= \cos^{-1}(P_0/V_0 I_0) < 0$. Other circuit quantities are found from:

$$R_c = \frac{E_1^2}{P_c} \tag{2.12}$$

$$I_c = \frac{P_c}{E_1} \tag{2.13}$$

$$I_m = \sqrt{I_0^2 - I_c^2} \tag{2.14}$$

$$X_m = \frac{E_1}{I_m} \tag{2.15}$$

$$a \approx \frac{V_0}{E_2} \tag{2.16}$$

Short-Circuit Test

In this test, one winding is short-circuited across its terminals, and a reduced voltage is applied to the other winding. This reduced voltage is of such a magnitude as to cause a specific value of current—usually, rated current—to flow in the short-circuited winding. Again, the choice of the winding to be short-circuited is usually determined by the measuring equipment available for use in the test. However, care must be taken to note which winding is short-circuited, for this determines the reference winding for expressing the impedance components obtained by this test. Let the secondary be short-circuited and the reduced voltage be applied to the primary.

With a very low voltage applied to the primary winding, the core-loss current and magnetizing current become very small, and the equivalent circuit reduces to that of Fig. 2-6. Thus, if P_s, I_s, and V_s are the input power, current, and voltage under short circuit, then, referred to the primary,

$$Z_s = \frac{V_s}{I_s} \tag{2.17}$$

$$R_1 + a^2 R_2 \equiv R_s = \frac{P_s}{I_s^2} \tag{2.18}$$

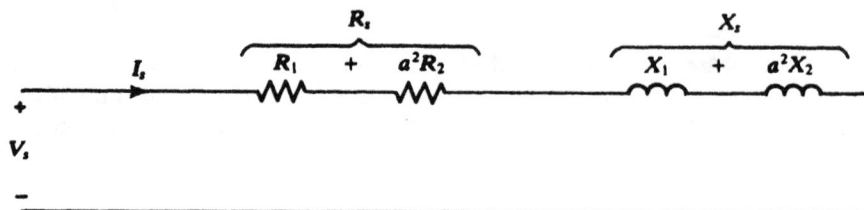

Fig. 2-6

$$X_1 + a^2 X_2 \equiv X_s = \sqrt{Z_s^2 - R_s^2} \qquad (2.19)$$

Given R_1 and a, R_2 can be found from (2.18). In (2.19) it is usually assumed that the leakage reactance is divided equally between the primary and the secondary; that is,

$$X_1 = a^2 X_2 = \frac{1}{2} X_s \qquad (2.20)$$

2.6 TRANSFORMER CONNECTIONS

Of the eight types of transformer connection shown in Table 2-1, the first six are for purposes of voltage transformation and the last two are for changing the number of phases. (Not included is the single-phase voltage transformer.) Each line segment in the diagrams corresponds to one winding of a two-winding transformer.

Table 2-1

Type of Connection	Primary	Secondary
Two-phase	 $\lfloor\_$	$\lfloor\_$ or $\lfloor$
Three-phase, delta-delta	 △	 △
Three-phase, delta-wye	 △	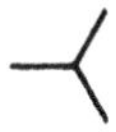 ⤙
Three-phase, wye-wye	 Y	 Y
Three-phase, open-delta	 ∠	 ∠
Three-phase, tee	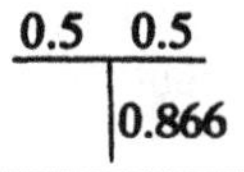 0.5 0.5 $\mid$0.866	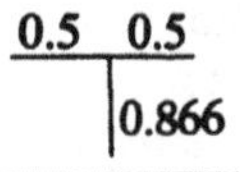 0.5 0.5 $\mid$0.866
Two-phase-to-three-phase (Scott)	$\lfloor\_$	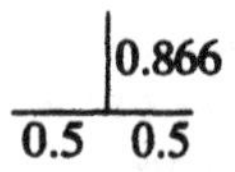 $\mid$0.866 / 0.5 0.5
Three-phase-to-six-phase (diametrical)	 Y or  ▷	 ✳

It is important to observe the polarity markings in polyphase transformer connections, and for the sake of illustration the connection of three identical transformers in delta-wye is shown in some detail in Fig. 2-7, which also shows the voltage phasor diagram. Notice the 30° phase shift between the line and phase voltages.

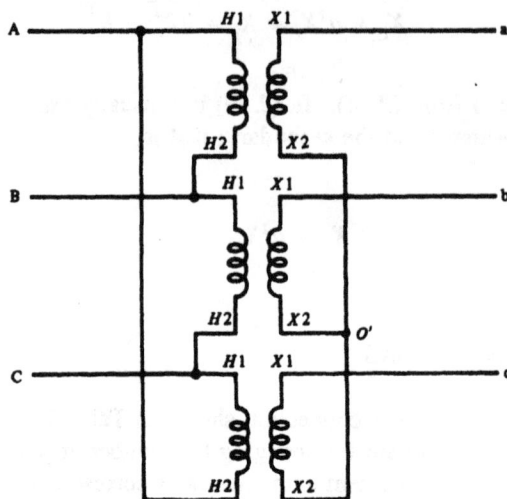

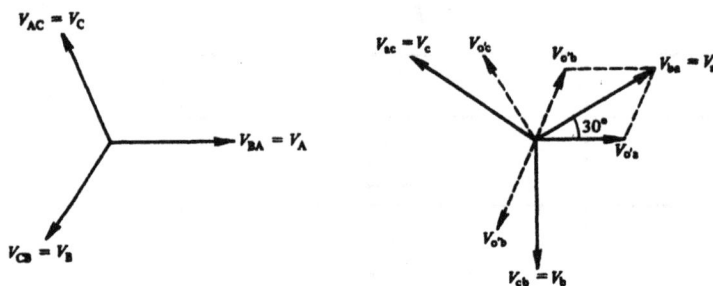

Fig. 2-7

2.7 AUTOTRANSFORMERS

An *autotransformer* is a single-winding transformer; it is a very useful device for some applications because of its simplicity and relatively low cost compared to multiwinding transformers. However, it does not provide electrical isolation and therefore cannot be used where this feature is required. The autotransformer circuit, Fig. 2-8, can be developed from a two-winding transformer by connecting the two windings electrically in series so that the polarities are additive. Assume that this has been done in the circuit of Fig. 2-8, where the primary of the two-winding transformer is winding *AB* and the secondary is winding *BC*. The primary of the autotransformer is now the sum of these two windings, *AC*, and the secondary is winding *BC*. Hence, the autotransformer voltage and turns ratio is

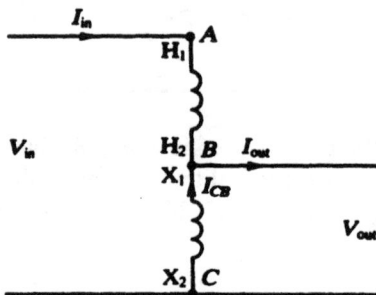

Fig. 2-8

$$a' = \frac{E_{AB} + E_{BC}}{E_{BC}} = \frac{N_{AB} + N_{BC}}{N_{BC}} = a + 1 \qquad (2.21)$$

where a is the voltage and turns ratio of the original two-winding transformer. Besides furnishing a greater transformation ratio, a pair of windings can also deliver more voltamperes (apparent power) when connected as an autotransformer than when connected as a two-winding transformer. The reason is that the transfer of voltamperes from primary to secondary in an autotransformer is not only by induction, as in a two-winding transformer, but by conduction as well.

2.8 INSTRUMENT TRANSFORMERS

Instrument transformers are of two kinds: current transformers (CTs) and potential transformers (PTs). These are used to supply power to ammeters, voltmeters, wattmeters, relays, and so on. Instrument transformers are used for (1) reducing the measured quantity to a low value which can be indicated by standard instruments (a standard voltmeter may be rated at 120 V and an ammeter at 5 A); and (2) isolating the instruments from high-voltage sources for safety. A connection diagram of a CT and a PT with an ammeter, a voltmeter, and a wattmeter is shown in Fig. 2-9(a). The load on the instrument transformer is called the *burden*. Depending on the burden, instrument transformers are rated from 25 to 500 VA. However, a PT or a CT is much (two to six times) bigger than a power transformer of the same rating.

An ideal instrument transformer has no phase difference between the primary and secondary voltages (or currents), which are independent of the burden. Like the ideal power transformer, the voltage ratio of an ideal PT is exactly equal to its turns ratio. The current ratio of an ideal CT is exactly equal to the inverse of the turns ratio. In practice, however, load-dependent ratio and phase-angle errors are present in instrument transformers.

The principle of operation of an instrument transformer is no different from that of an ordinary power transformer. Thus they have similar phasor diagrams, as shown in Fig. 2-9(b). It is clear from this diagram that the secondary impedance drop causes a phase displacement α, and the primary impedance drop a phase displacement β; the exciting current I_0 causes a further phase displacement γ, so that the angle between the primary voltage and current is $(\theta_2 + \alpha + \beta + \gamma)$, compared with an angle θ_2 between the secondary voltage and current. Thus the transformer introduces a phase-angle error $(\alpha + \beta + \gamma)$. Moreover, V_1 and V_2 will be only approximately in the ratio of the number of turns. In order to nullify or reduce the errors, instrument transformers are designed with (1) small leakage reactances and low resistances which reduce angles α and β; (2) low flux densities and good transformer iron, which reduces the exciting current I_0 and therefore angle γ; and (3) less than a nominal turns ratio, which compensates for the ratio error. For a constant burden, the instruments may be calibrated, or corrected, against the load.

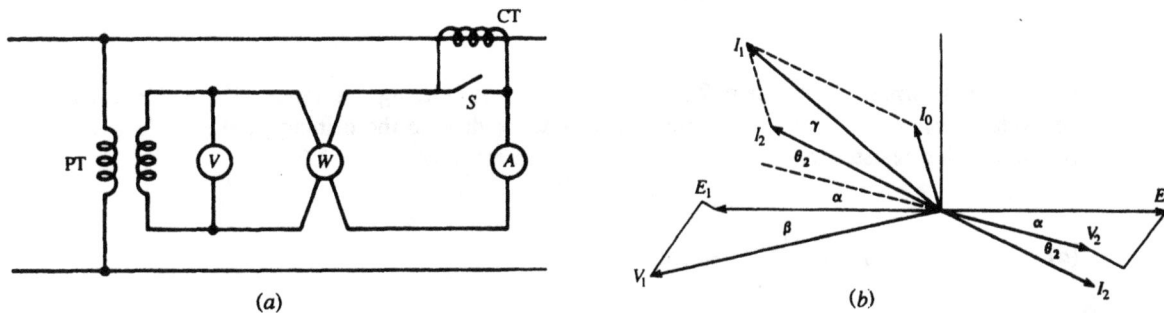

Fig. 2-9

Solved Problems

2.1.   The primary of a transformer has 200 turns and is excited by a 60-Hz, 200-V source. What is the maximum value of the core flux?

From the emf equation, (2.9),

$$\phi_m = \frac{E_1}{4.44 f N_1} = \frac{220}{(4.44)(60)(200)} = 4.13 \text{ mWb}$$

2.2.   A voltage $v = 155.5 \sin 377t + 15.5 \sin 1131t$ (V) is applied to the primary of the transformer of Problem 2.1. Neglecting leakage, determine the instantaneous and rms values of the core flux.

From (2.2),

$$\phi = \frac{1}{200} \int (155.5 \sin 377t + 15.5 \sin 1131t) \, dt$$
$$= -2.05 \cos 377t - 0.068 \cos 1131t \quad (\text{mWb})$$

The two components of ϕ have frequencies in integral ratio (1:3). Hence their separate rms values, $2.05/\sqrt{2}$ and $0.068/\sqrt{2}$, combine as follows:

$$\phi_{rms} = \sqrt{\left(\frac{2.05}{\sqrt{2}}\right)^2 + \left(\frac{0.068}{\sqrt{2}}\right)^2} = 1.45 \text{ mWb}$$

2.3.   A 60-Hz transformer having a 480-turn primary winding takes 80 W in power and 1.4 A in current at an input voltage of 120 V. If the primary winding resistance is 0.25 Ω, determine (a) the core loss, (b) the no-load power factor, and (c) the maximum core flux (neglect the primary resistance and reactance drops).

(a) $P_c = 80 - (1.4)^2(0.25) = 79.5$ W

(b) $\cos \theta_0 = \dfrac{80}{(1.4)(120)} = 0.476$

(c) $\phi_m = \dfrac{120}{(4.44)(60)(480)} = 0.94$ mWb

2.4.   For the transformer of Problem 2.3, evaluate the magnetizing reactance, X_m, and the core-loss equivalent resistance, R_c (a) neglecting the impedance drop in the primary; (b) including the effect of the winding resistance, $R_1 = 0.25$ Ω, and leakage reactance, $X_1 = 1.2$ Ω.

(a) $R_c = \dfrac{(120)^2}{80} = 180 \ \Omega$

$$I_c = \frac{120}{180} = 0.67 \text{ A}$$

$$I_m = \sqrt{(1.4)^2 - (0.67)^2} = 1.23 \text{ A}$$

$$X_m = \frac{120}{1.23} = 97.5 \text{ } \Omega$$

(b)   From Problem 2.3(b), $\theta_0 = \cos^{-1} 0.476 = -61.6°$. Then, by (2.11),

$$\mathbf{E}_1 = 120\angle 0° - (1.4\angle -61.6°)(0.25 + j1.25) \quad \text{or} \quad E_1 \approx 118.29 \text{ V}$$

and we have:

$$R_c = \frac{(118.29)^2}{79.5} = 176 \text{ } \Omega$$

$$I_c = \frac{118.29}{176} = 0.672 \text{ A}$$

$$I_m = \sqrt{(1.4)^2 - (0.672)^2} = 1.228 \text{ A}$$

$$X_m = \frac{118.29}{1.228} = 96.3 \text{ } \Omega$$

2.5. The parameters of the equivalent circuit of a 150-kVA, 2400-V/240-V transformer, shown in Fig. 2-3, are $R_1 = 0.2$ Ω, $R_2 = 2$ mΩ, $X_1 = 0.45$ Ω, $X_2 = 4.5$ mΩ, $R_c = 10$ kΩ, and $X_m = 1.55$ kΩ. Using the circuit referred to the primary, determine the (a) *voltage regulation* and (b) *efficiency* of the transformer operating at rated load with 0.8 lagging power factor.

See Figs. 2-4(a) and 2-5. Given $V_2 = 240$ V, $a = 10$, and $\theta_2 = \cos^{-1} 0.8 = -36.8°$,

$$a\mathbf{V}_2 = 2400\angle 0° \text{ V}$$

$$I_2 = \frac{150 \times 10^3}{240} = 625 \text{ A} \quad \text{and} \quad \frac{\mathbf{I}_2}{a} = 62.5\angle -36.8° = 50 - j37.5 \text{ A}$$

Also, $a^2 R_2 = 0.2$ Ω and $a^2 X_2 = 0.45$ Ω, so that

$$\mathbf{E}_1 = (2400 + j0) + (50 - j37.5)(0.2 + j0.45)$$

$$= 2427 + j15 = 2427\angle 0.35° \text{ V}$$

$$\mathbf{I}_m = \frac{2427\angle 0.35°}{1550\angle 90°} = 1.56 \angle -89.65° = 0.0095 - j1.56 \text{ A}$$

$$\mathbf{I}_c = \frac{2427 + j15}{10 \times 10^3} \approx 0.2427 + j0 \text{ A}$$

Therefore

$$\mathbf{I}_0 = \mathbf{I}_c + \mathbf{I}_m = 0.25 - j1.56 \text{ A}$$

$$\mathbf{I}_1 = \mathbf{I}_0 + (\mathbf{I}_2/a) = 50.25 - j39.06 = 63.65\angle -37.85° \text{ A}$$

$$\mathbf{V}_1 = (2427 + j15) + (50.25 - j39.06)(0.2 + j0.45)$$

$$= 2455 + j30 = 2455\angle 0.7° \text{ V}$$

(*a*) percent regulation $\equiv \dfrac{V_{\text{no-load}} - V_{\text{load}}}{V_{\text{load}}} \times 100$

$$= \dfrac{V_1 - aV_2}{aV_2} \times 100 = \dfrac{2455 - 2400}{2400} \times 100 = 2.3\%$$

(*b*) efficiency $\equiv \dfrac{\text{output}}{\text{input}} = \dfrac{\text{output}}{\text{output} + \text{losses}}$

output $= (150 \times 10^3)(0.8) = 120$ kW

losses $= I_1^2 R_1 + I_c^2 R_c + I_2^2 R_2$

$= (63.65)^2(0.2) + (0.2427)^2(10 \times 10^3) + (625)^2(2 \times 10^{-3}) = 2.18$ kW

Hence   efficiency $= \dfrac{120}{122.18} = 0.982 = 98.2\%$

2.6.   Corresponding to Fig. 2-4(*a*), an approximate equivalent circuit of a transformer is shown in Fig. 2-10. Using this circuit, repeat the calculations of Problem 2.5 and compare the results. Draw a phasor diagram showing all the voltages and currents of the circuit of Fig. 2-10.

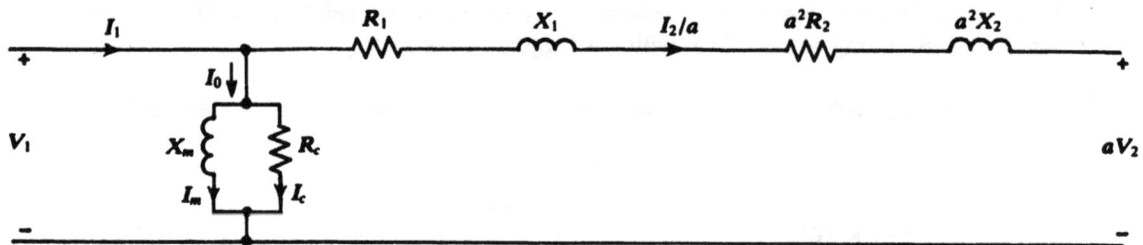

Fig. 2-10

From Problem 2.5,

$$aV_2 = 2400\angle 0° \quad V$$

$$\dfrac{I_2}{a} = 50 - j37.5 \quad A$$

$$R_1 + a^2 R_2 = 0.4 \ \Omega$$

$$X_1 + a^2 X_2 = 0.9 \ \Omega$$

Hence

$$V_1 = (2400 + j0) + (50 - j37.5)(0.4 + j0.9)$$

$$= 2453 + j30 = 2453\angle 0.7° \quad V$$

$$I_c = \dfrac{2453\angle 0.7°}{10 \times 10^3} = 0.2453\angle 0.7 \quad A$$

$$I_m = \dfrac{2453\angle 0.7°}{1550\angle 90°} = 1.58\angle -89.3° \quad A$$

$$I_0 = 0.2453 - j1.58 \quad A$$

$$I_1 = 50.25 - j39.08 = 63.66 \angle{-37.9°} \quad A$$

The phasor diagram is shown in Fig. 2-11.

(a) $$\text{percent regulation} = \frac{2453 - 2400}{2400} \times 100 = 2.2\%$$

(b) efficiency $$= \frac{120 \times 10^3}{120 \times 10^3 + (63.66)^2\,(0.4) + (0.2453)^2(10 \times 10^3)} = 0.982 = 98.2\%$$

Notice that the approximate circuit yields results that are sufficiently accurate.

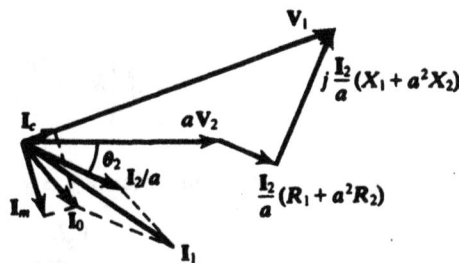

Fig. 2-11

2.7. The ohmic values of the circuit parameters of a transformer, having a turns ratio of 5, are $R_1 = 0.5$ Ω; $R_2 = 0.021\ \Omega$; $X_1 = 3.2\ \Omega$; $X_2 = 0.12\ \Omega$; $R_c = 350\ \Omega$, referred to the primary; and $X_m = 98\ \Omega$, referred to the primary. Draw the approximate equivalent circuits of the transformer, referred to (a) the primary and (b) the secondary. Show the numerical values of the circuit parameters.
   The circuits are respectively shown in Fig. 2-12(a) and Fig. 2-12(b). The calculations are as follows:

(a) $R' \equiv R_1 + a^2 R_2 = 0.5 + (5)^2(0.021) = 1.025\ \Omega$

$X' \equiv X_1 + a^2 X_2 = 3.2 + (5)^2(0.12) = 6.2\ \Omega$

$R_c' = 350\ \Omega$

$X_m' = 98\ \Omega$

(b) $R'' \equiv \dfrac{R_1}{a^2} + R_2 = \dfrac{0.5}{25} + 0.021 = 0.041\ \Omega$

$X'' \equiv \dfrac{X_1}{a^2} + X_2 = \dfrac{3.2}{25} + 0.12 = 0.248\ \Omega$

$R_c'' = \dfrac{350}{25} = 14\ \Omega$

$X_m'' = \dfrac{98}{25} = 3.92\ \Omega$

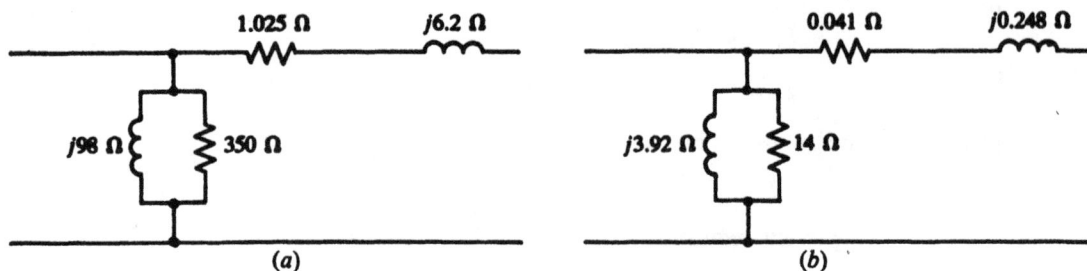

Fig. 2-12

2.8. Using the approximate circuit of Fig. 2-10, determine the secondary current at which the transformer will have a maximum efficiency.

Let

$$P_{core} \equiv I_c^2 R_c \quad \text{and} \quad P_{copper} \equiv I_2^2 \left(R_2 + \frac{R_1}{a^2} \right)$$

denote the core and copper losses, respectively. The efficiency is given by

$$\eta \equiv \frac{V_2 I_2}{V_2 I_2 + P_{core} + P_{copper}} = \frac{V_2 I_2}{V_2 I_2 + P_{core} + I_2^2 \left(R_2 + (R_1/a^2) \right)}$$

For η to be a maximum,

$$\frac{d\eta}{dI_2} = 0$$

which, under the assumption that P_{core} is independent of I_2, implies that

$$\left[V_2 I_2 + P_{core} + I_2^2 \left(R_2 + \frac{R_1}{a^2} \right) \right] V_2 - V_2 I_2 \left[V_2 + 2I_2 \left(R_2 + \frac{R_1}{a^2} \right) \right] = 0$$

or

$$P_{core} - P_{copper} = 0$$

Hence, the maximum efficiency is at the load for which the copper loss equals the core loss. The maximizing current I_2 is given by

$$I_2 = \left[\frac{P_{core}}{R_2 + (R_1/a^2)} \right]^{1/2}$$

2.9. A 110-kVA, 2200-V/110-V, 60-Hz transformer has the following circuit constants: $R_1 = 0.22 \ \Omega$, $R_2 = 0.5 \ \text{m}\Omega$, $X_1 = 2.0 \ \Omega$, $X_2 = 5 \ \text{m}\Omega$, $R_c = 5.5 \ \text{k}\Omega$, and $X_m = 1.1 \ \text{k}\Omega$. During one day (24 hours) the transformer has the following load cycle: 4 h on no-load; 8 h on ¼ full-load at 0.8 power factor; 8 h on ½ full-load at unity power factor; and 4 h on full-load at unity power factor. Assuming a constant core loss of 1.346 kW, find the all-day efficiency of the transformer.

$$\text{all-day efficiency} \equiv \frac{\text{energy output for 24 h}}{\text{energy input for 24 h}}$$

$$\text{output for 24 h} = \left(4 \times 0 + 8 \times \frac{1}{4} \times 0.8 + 8 \times \frac{1}{2} \times 1 + 4 \times 1 \times 1\right) 110 = 1056 \text{ kWh}$$

The total core loss for 24 h is $(24)(1.346 \times 10^3) = 32.3$ kWh. Determining the secondary and primary currents during the different periods as in Problem 2.5, we calculate the following ohmic losses in the windings:

for 8 h on $\frac{1}{4}$ full-load: $[(250)^2(5 \times 10^{-3}) + (14.1)^2(0.22)]8 = 2.85$ kWh

for 4 h on $\frac{1}{2}$ full-load: $[(500)^2(5 \times 10^{-3}) + (26.6)^2(0.22)]4 = 5.62$ kWh

for 8 h on full-load: $[(1000)^2(5 \times 10^{-3}) + (51.7)^2(0.22)]8 = 44.70$ kWh

Then the total ohmic loss for 24 h is 53.17 kWh, and

$$\eta_{\text{all-day}} = \frac{1056}{1056 + 32.3 + 53.17} = 0.925 = 92.5\%$$

2.10. The results of open-circuit and short-circuit tests on a 25-kVA, 440-V/220-V, 60-Hz transformer are as follows:

Open-circuit test. Primary open-circuited, with instrumentation on the low-voltage side. Input voltage, 220 V; input current, 9.6 A; input power, 710 W.

Short-circuit test. Secondary short-circuited, with instrumentation on the high-voltage side. Input voltage, 42 V; input current, 57 A; input power, 1030 W.

Obtain the parameters of the exact equivalent circuit (Fig. 2-4), referred to the high-voltage side. Assume that $R_1 = a^2 R_2$ and $X_1 = a^2 X_2$.

From the short-circuit test:

$$Z_{s1} = \frac{42}{57} = 0.737 \ \Omega$$

$$R_{s1} = \frac{1030}{(57)^2} = 0.317 \ \Omega$$

$$X_{s1} = \sqrt{(0.737)^2 - (0.317)^2} = 0.665 \ \Omega$$

Consequently,

$$R_1 = a^2 R_2 = 0.158 \ \Omega \qquad R_2 = 0.0395 \ \Omega$$

$$X_1 = a^2 X_2 = 0.333 \ \Omega \qquad X_2 = 0.0832 \ \Omega$$

From the open-circuit test:

$$\theta_0 = \cos^{-1} \frac{710}{(9.6)(220)} = \cos^{-1} 0.336 = -70°$$

$$\mathbf{E}_2 = 220\angle 0° - (9.6\angle -70°)(0.0395 + j0.0832) \approx 219\angle 0° \text{ V}$$

$$P_{c2} = 710 - (9.6)^2(0.0395) \approx 710 \text{ W} \quad (706.3 \text{ W, exact})$$

$$R_{c2} = \frac{(219)^2}{710} = 67.5 \ \Omega$$

$$I_{c2} = \frac{219}{67.5} = 3.24 \text{ A}$$

$$I_{m2} = \sqrt{(9.6)^2 - (3.24)^2} = 9.03 \text{ A}$$

$$X_{m2} = \frac{219}{9.03} = 24.24 \ \Omega$$

$$X_{m1} = a^2 X_{m2} = 97 \ \Omega$$

$$R_{c1} = a^2 R_{c2} = 270 \ \Omega$$

Thus, the equivalent circuit has the parameters as labeled in Fig. 2-13.

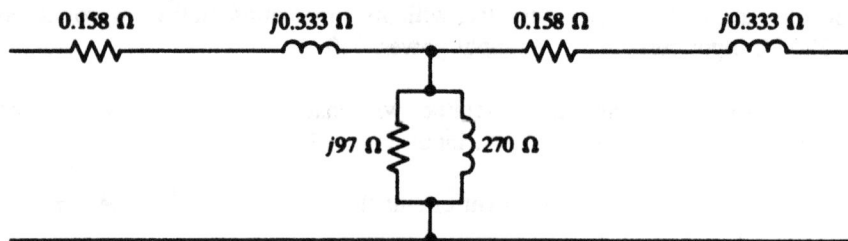

Fig. 2-13

2.11.   From the test data of Problem 2.10, obtain the values of the circuit constants for the approximate equivalent circuit referred to the low-voltage side.

The circuit has the appearance of Fig. 2-12(b), but now:

$$R_{c2} = \frac{(220)^2}{710} = 68.2 \ \Omega$$

$$I_{c2} = \frac{220}{68.2} = 3.22 \text{ A}$$

$$I_{m2} = \sqrt{(9.6)^2 - (3.22)^2} = 9.04 \text{ A}$$

$$X_{m2} = \frac{220}{9.04} = 24.33 \ \Omega$$

The values calculated from the short-circuit test data (in Problem 2.10) should all be referred to the secondary. Thus,

$$R_{s2} = \frac{0.317}{4} = 0.079 \ \Omega$$

$$X_{s2} = \frac{0.665}{4} = 0.166 \ \Omega$$

2.12. A two-winding transformer, with the windings identified as H_1H_2 and X_1X_2, is reconnected as an autotransformer as shown in Fig. 2-9. Compare approximately the voltage and voltamperage ratings of the autotransformer with those of the original two-winding transformer.

From Fig. 2-9, $V_{in} \approx E_{AB} + E_{BC} \approx E_{AB} + V_{out}$. Then,

$$b \equiv \text{autotransformer voltage ratio} = \frac{V_{in}}{V_{out}} \approx \frac{E_{AB}}{E_{BC}} + 1$$

$$a \equiv \text{two-winding voltage ratio} = \frac{E_{AB}}{E_{BC}}$$

Therefore,

$$\frac{b}{a} \approx 1 + \frac{E_{BC}}{E_{AB}} = 1 + \frac{N_{BC}}{N_{AB}}$$

Moreover,

$$(VA)_a \equiv \text{transformer voltampere rating} = V_{in}I_{in} \approx (E_{AB} + E_{BC})I_{in}$$

$$(VA)_t \equiv \text{two-winding voltampere rating} = E_{AB}I_{in}$$

so that

$$\frac{(VA)_a}{(VA)_t} \approx 1 + \frac{E_{BC}}{E_{AB}} \approx \frac{b}{a}$$

Thus, approximately, both voltage transformation and voltamperage rating are increased by the same factor when a two-winding transformer is reconnected as an autotransformer. This factor was denoted a'/a in (2.21).

2.13. Two transformers, with equivalent impedances $\mathbf{Z}'_e$ and $\mathbf{Z}''_e$ referred to the respective primaries, operate in parallel at a secondary terminal voltage $\mathbf{V}_t$ and a primary terminal voltage $\mathbf{V}_1$. The transformers have a' and a'' as their respective turns ratios. If the total primary current is $\mathbf{I}_1$, determine how the load is shared by the two transformers. Neglect the core losses and magnetizing current.

The equivalent circuits of two transformers in parallel are shown in Fig. 2-14, for which the following relations hold:

$$\mathbf{V}_1 = \mathbf{V}'_1 = a'\mathbf{V}_t + \mathbf{I}'_1\mathbf{Z}'_e$$

$$\mathbf{V}_1 = \mathbf{V}''_1 = a''\mathbf{V}_t + \mathbf{I}''_1\mathbf{Z}''_e$$

$$\mathbf{I}_1 = \mathbf{I}'_1 + \mathbf{I}''_1$$

Subtracting the first two equations, and then solving simultaneously with the third, we obtain for the two load currents:

$$\mathbf{I}_1' = \frac{-\mathbf{V}_t(a' - a'') + \mathbf{I}_1\mathbf{Z}_e''}{\mathbf{Z}_e' + \mathbf{Z}_e''} \qquad \mathbf{I}_1'' = \frac{\mathbf{V}_t(a' - a'') + \mathbf{I}_1\mathbf{Z}_e'}{\mathbf{Z}_e' + \mathbf{Z}_e''}$$

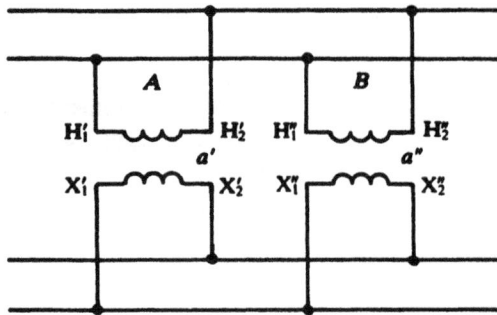

(a) Two transformers in parallel

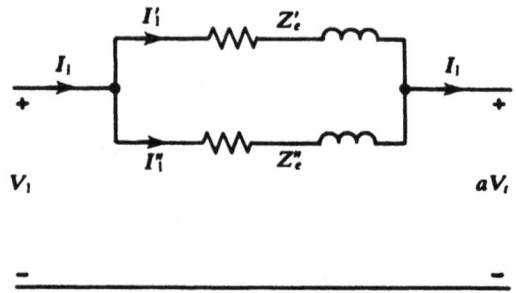

(b) Equivalent circuit for (a)

Fig. 2-14

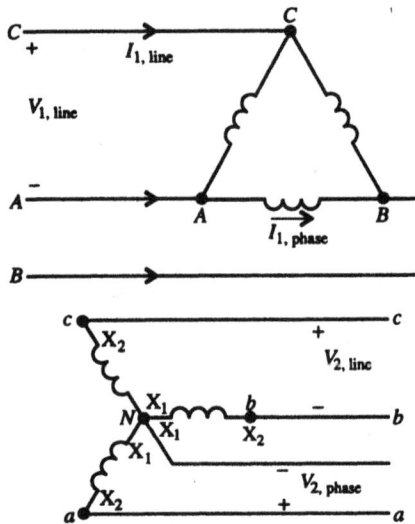

Fig. 2-15

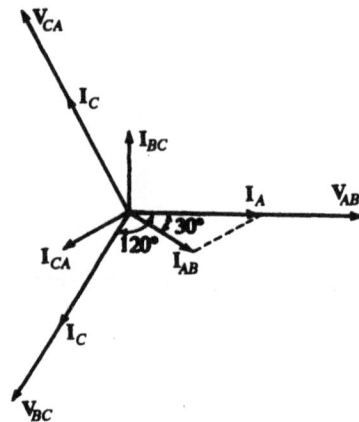

Fig. 2-16

2.14. Three single-phase transformers are connected to make a delta-wye, three-phase transformer, as shown in Fig. 2-15. Draw a phasor diagram showing all the voltages and currents in the transformers.

See Fig. 2-16.

2.15. Two transformers, each rated at 100 kVA, 11000 V/2300 V, 60 Hz, are connected in open delta on primary and secondary sides. (a) What is the total load that can be supplied from this transformer bank? (b) A 120-kVA, 2300-V, 0.866-lagging-power-factor, three-phase, delta-connected load is connected to the transformer bank. What is the line current on the high-voltage side?

(a)   load on open delta = $\sqrt{3}$ × (kVA-rating of each transformer)

$$= \sqrt{3} \times 100 = 173.2 \text{ kVA}$$

The circuit and phasor diagrams are shown in Fig. 2-17.

(b) For the delta-connected load:

$$I_{AB} = I_{BC} = I_{CA} = \frac{1}{3}\left(\frac{120 \times 10^3}{2300}\right) = 17.4 \text{ A}$$

From the phasor diagram, Fig. 2-17(b),

$$\mathbf{I}_A = \mathbf{I}_{AB} - \mathbf{I}_{CA} = \left(2 \times \frac{\sqrt{3}}{2} \times 17.4\right)\angle 0° = 30.12\angle 0° \text{ A}$$

$$\text{transformation ratio } a = \frac{11000}{2300} = 4.78$$

$$I_{line} \equiv \text{current in the } 11000\text{-V winding} = \frac{30.12}{4.78} = 6.3 \text{ A}$$

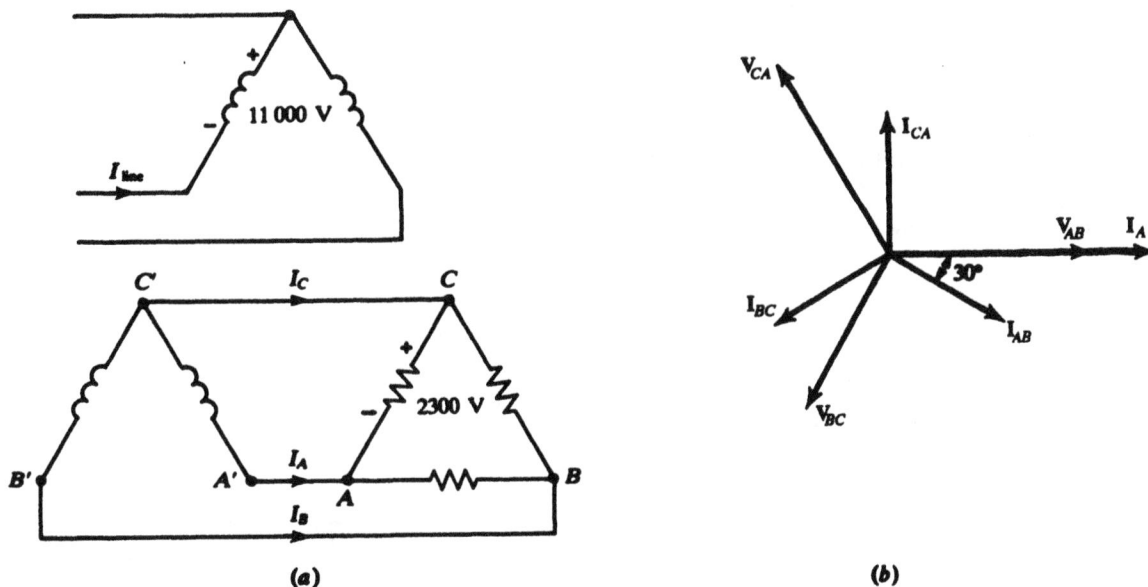

(a)   (b)

Fig. 2-17

2.16. A 25-Hz, 120-V/30-V, 500-VA transformer is to be used on a 60-Hz source. If the core flux density is to remain unchanged, determine (a) the maximum permissible primary voltage, and (b) the new (60-Hz) rated secondary voltage and current.

(a) By (2.9), the primary voltage will vary directly with frequency. Hence,

$$\text{maximum primary voltage} = \frac{60}{25}(120) = 288 \text{ V}$$

(b)

$$\text{rated } V_2 = \frac{60}{25}(30) = 72 \text{ V}$$

$$\text{rated } I_2 = \frac{500}{30} = 16.67 \text{ A} \quad \text{(same as at 25 Hz)}$$

2.17. When a certain transformer is connected to rated sinusoidal voltage of 115 V at rated frequency, it draws an exciting current that has the following rms components: fundamental, 2 A; third harmonic, 0.8 A; fifth harmonic, 0.5 A. This transformer and two others identical to it are connected in wye to a balanced, 4-wire, 3-phase source of rated frequency with 115 V from line to neutral. Compute the rms values of line and neutral currents.

$$I_{\text{line}} = \sqrt{(2)^2 + (0.8)^2 + (0.5)^2} = 2.21 \text{ A}$$

By drawing the waveforms of the three phase currents, each decomposed into its three harmonics, it may be verified that the third-harmonic currents are all in phase, whereas the fundamentals and fifth-harmonics cancel. Consequently,

$$I_{\text{neutral}} = 3(0.8) = 2.4 \text{ A}$$

2.18. A quantity is expressed in *per unit* if it is divided by a chosen *base quantity* (having the same physical dimension). Suppose that for a 10-kVA, 2400-V/240-V transformer we choose

$$P_{\text{base}} = 10 \text{ kW} \qquad V_{1,\text{base}} = 2400 \text{ V} \qquad V_{2,\text{base}} = 240 \text{ V}$$

This transformer has the following test data:

open-circuit test (on low-voltage side):   240 V, 0.8 A, 80 W
short-circuit test (on high-voltage side):   80 V, 5.1 A, 220 W

Convert all test data into per-unit values and find the series equivalent resistance in per unit.

$$I_{1,\text{base}} = \frac{10 \times 10^3}{2400} = 4.17 \text{A} \qquad I_{2,\text{base}} = 41.7 \text{ A}$$

In per unit (pu), the open-circuit data are

$$V_0 = \frac{240}{240} = 1 \text{ pu} \qquad I_0 = \frac{0.8}{41.7} = 0.019 \text{ pu} \qquad P_0 = \frac{80}{10 \times 10^3} = 0.008 \text{ pu}$$

and the short-circuit data are

$$V_s = \frac{80}{2400} = 0.0333 \text{ pu} \qquad I_s = \frac{5.1}{4.17} = 1.22 \text{ pu} \qquad P_s = \frac{220}{10 \times 10^3} = 0.022 \text{ pu}$$

The equivalent impedance and power factor are

$$Z_e = \frac{V_s \text{ (pu)}}{I_s \text{ (pu)}} = \frac{0.0333}{1.22} = 0.0273 \text{ pu}$$

$$\cos \theta_e = \frac{P_s}{V_s I_s} = \frac{0.022}{(0.0333)(1.22)} = 0.54$$

and so $R_e = Z_e \cos \theta_e = 0.0148$ pu.

2.19. A 75-kVA, 230-V/115-V, 60-Hz transformer was tested with the following results:

short-circuit test: 9.5 V, 326 A, 1200 W

open-circuit test: 115 V, 16.3 A, 750 W

Determine the (*a*) equivalent impedance in high-voltage terms; (*b*) equivalent impedance in per unit; (*c*) voltage regulation at rated load, 0.8 power factor lagging; (*d*) efficiency at rated load, 0.8 power factor lagging, and at 1/2 load, unity power factor; (*e*) maximum efficiency and the current at which it occurs.

(*a*)
$$Z_s = \frac{9.5}{326} = 0.029 \ \Omega$$

(*b*) Proceeding as in Problem 2.18,

$$\text{per-unit } V_s = \frac{9.5}{230} = 0.0413 \ \text{pu}$$

$$\text{per-unit } I_s = \frac{326}{326} = 1 \ \text{pu}$$

$$\text{per-unit } Z_s = \frac{V_s \ (\text{pu})}{I_s \ (\text{pu})} = \frac{0.0413}{1} = 0.0413 \ \text{pu}$$

(*c*)
$$\text{per-unit } P_s = \frac{1200}{75 \times 10^3} = 0.016 \ \text{pu} = I_{\text{pu}}^2 R_{\text{pu}}$$

Thus,

$$R_{\text{pu}} = 0.016 \quad X_{\text{pu}} = \sqrt{(0.0413)^2 - (0.016)^2} = 0.0381 \ \text{pu}$$

$$\mathbf{V_0} = \mathbf{V} + \mathbf{IZ} = 1 + (0.8 - j0.6)(0.016 + j0.0381)$$

whence $V_0 = 1.036$ pu. Then,

$$\text{voltage regulation} = \frac{V_0 - V_2}{V_2} = \frac{1.036 - 1}{1} = 0.036 \ \text{pu} = 3.6\%$$

(*d*)
$$\eta_{\text{rated load}} = \frac{(75 \times 10^3)(0.8)}{60 \times 10^3 + 1200 + 750} = 96.85\%$$

$$\eta_{1/2 \text{ rated load}} = \frac{(37.5 \times 10^3)(1)}{37.5 \times 10^3 + 300 + 750} = 97.27\%$$

(*e*) According to Problem 2.8, the maximizing current, I_1, is given by

$$I_1^2 R_e \equiv \text{copper loss} = \text{core loss} \approx 750 \ \text{W}$$

The short-circuit test gives the equivalent resistance, R_e, as

$$R_e = \frac{1200}{(326)^2} \ \Omega$$

Hence,

$$I_1 = 326 \sqrt{\frac{750}{1200}} = (326)(0.79) = 257.72 \text{ A}$$

The power output is

$$\frac{I_1}{326} (75 \times 10^3) = (0.79)(75 \times 10^3) \text{ W}$$

and so

$$\eta_{max} = \frac{(0.79)(75 \times 10^3)}{(0.79)(75 \times 10^3) + 750 + 750} = 97.53\%$$

Supplementary Problems

2.20. The primary of an ideal transformer has 1000 turns, and is rated at 220 V 60 Hz. If the core cross section is 10 cm^2, what is the operating flux density? *Ans.* 0.826 T

2.21. The *B-H* curve of the core of a transformer is as shown in Fig. 1-17(*b*), and the maximum flux density is 1.2 T for a sinusoidal input voltage. Show qualitatively that the exciting current is nonsinusoidal.

2.22. A flux, $\phi = 2 \sin 377t + 0.08 \sin 1885t$ (mWb), completely links a 500-turn coil. Calculate the (*a*) instantaneous and (*b*) rms induced voltage in the coil.
Ans. (*a*) $v = 377 \cos 377t + 75.4 \cos 1885t$ (V); (*b*) $V = 271.86$ V

2.23. A 100-kVA, 60-Hz, 2200-V/220-V transformer is designed to operate at a maximum flux density of 1 T and an induced voltage of 15 volts per turn. Determine the number of turns of (*a*) the primary winding, (*b*) the secondary winding. (*c*) What is the cross-sectional area of the core?
Ans. (*a*) 147 turns; (*b*) 15 turns; (*c*) 0.0563 m^2

2.24. A transformer has a turns ratio of 5. (*a*) If a 100-Ω resistor is connected across the secondary, what is its resistance referred to the primary? (*b*) If the same resistor is instead connected across the primary, what is its resistance referred to the secondary? *Ans.* (*a*) 2500 Ω; (*b*) 4 Ω

2.25. Refer to Fig. 2-1, and let the core have a reluctance $\mathfrak{R}$. A resistance R is connected across the secondary. The core flux is sinusoidal, of frequency ω, and the turns ratio is N_1/N_2. Derive an expression for the instantaneous primary current in terms of N_1, N_2, ω, R, $\mathfrak{R}$, and the primary induced voltage E_1.
Ans. $i_1 = (N_2/N_1)^2 E_1/R \cos \omega t + (E_1 \mathfrak{R}/\omega N_1^2) \sin \omega t$. (Note that $\omega N_1^2/\mathfrak{R} \equiv X_m$, the magnetizing reactance of the transformer.)

2.26. Repeat Problem 2.5, but with the equivalent circuit referred to the secondary. Draw the phasor diagram and verify that the percent voltage regulation and efficiency are consistent with the values previously found.

2.27. Refer to Fig. 2-4. For a 110-kVA, 2200-V/110-V transformer the ohmic values of the circuit parameters

are $R_1 = 0.22\ \Omega$, $R_2 = 0.5\ m\Omega$, $X_1 = 2.0\ \Omega$, $X_2 = 5\ m\Omega$, $R_c = 5494.5\ \Omega$, and $X_m = 1099\ \Omega$. Calculate (a) the voltage regulation and (b) the efficiency of the transformer, at full-load and unity power factor.
Ans. (a) 1.53%; (b) 98.3%

2.28. Repeat 2.27, but with the approximate equivalent circuit referred to the secondary. Draw the phasor diagram.

2.29. Find the core losses of the transformer of Problem 2.27. *Ans.* 890.5 W

2.30. An *ideal* 220/110-V transformer carries a $(6 + j8)$ - Ω load at 110 V. Under this condition, calculate the input (a) volt-amp; (b) power (in watt); (c) power factor; and (d) impedance (all referred to the 220-V side). *Ans.* (a) 2420 VA; (b) 726 W; (c) 0.6 lagging; (d) 40 Ω

2.31. Open-circuit and short-circuit tests are performed on a 10-kVA, 220-V/110-V, 60-Hz transformer. Both tests are performed with instrumentation on the high-voltage side, and the following data are obtained:

open-circuit test: input power, 500 W; input voltage, 220 V; input current, 3.16 A

short-circuit test: input power, 400 W; input voltage, 65 V; input current, 10 A

Determine the parameters of the approximate equivalent circuit referred to the (a) primary and (b) secondary.
Ans. (a) $R_c = 96.8\ \Omega$, $X_m = 100\ \Omega$, $R_1 + a^2R_2 = 4\ \Omega$, $X_1 + a^2X_2 = 5.1\ \Omega$;
(b) $R_c = 24.2\ \Omega$, $X_m = 25\ \Omega$, $R_2 + (R_1/a^2) = 1\ \Omega$, $X_2 + (X_1/a^2) = 1.28\ \Omega$

2.32. The transformer represented by the circuit shown in Fig. 2-12(a) supplies a 5-kVA load at 440 V and 0.8 leading power factor. Calculate the total reactive kVA input to the transformer. *Ans.* −0.468 kVAr

2.33. Repeat Problem 2.32 for 0.8 lagging power factor. *Ans.* 6.285 kVAr

2.34. If an autotransformer is made from a two-winding transformer having a turns ratio $N_1/N_2 = a$, show that:

$$\frac{\text{magnetizing current as an autotransformer}}{\text{magnetizing current as a 2-winding transformer}} = \frac{a-1}{a}$$

$$\frac{\text{short-circuit current as an autotransformer}}{\text{short-circuit current as a 2-winding transformer}} = \frac{a}{a-1}$$

2.35. Two transformers, operating in parallel, deliver a 230-V, 400-kVA load at 0.8 power factor lagging. One transformer is rated at 2300 V/230 V and has an impedance of 1.84∠84.2° Ω, referred to the primary. The corresponding data for the second transformer are 2300 V/225 V and 0.77∠82.5° Ω. Calculate (a) the current and (b) the power delivered by each transformer.
Ans. (a) 526 A, 1212 A; (b) 93.35 kW, 226.65 kW

2.36. If the load on the transformers in Problem 2.31 is taken off completely, determine the no-load (or circulating) current of the transformers. *Ans.* 19.58 A

2.37. Use the data of Problem 2.35 and assume that both transformers have the same transformation ratio. For

this case, compute (a) the circulating current and (b) the current supplied by each transformer.
Ans. (a) 0 A;   (b) 513 A, 1226 A

2.38. The high-voltage windings of three 100-kVA, 19000-V/220-V transformers are connected in delta. The phase windings carry rated current at 0.866 lagging power factor. Determine the primary line and phase voltages and currents.   *Ans.* $\mathbf{V}_{AB} = 19000\angle 0°$ V, $\mathbf{V}_{BC} = V_{AB}\angle 120°$, $\mathbf{V}_{CA} = V_{AB}\angle 240°$,
$\mathbf{I}_{AB} = 5.26\angle -30°$ A, $\mathbf{I}_A = 9.1\angle 0°$ A

2.39. The flux linking a 500-turn coil is given by

$$\phi = 8t^2 \quad \text{(Wb)}$$

where t is in s. Plot the induced voltage versus t. Compute the induced voltage at $t = 2$ s and at $t = 4$ s.
Ans. 16 kV; 32 kV

2.40. Bearing in mind the equivalent circuit of a transformer referred to the primary, as shown in Fig. 2-3(a), answer the following questions. (a) What experimental tests are used in obtaining the impedances shown? (b) What approximations are involved in relating the results of the experimental tests to the impedances shown? (c) Where does leakage flux in the transformer core show up in the diagram? (d) Which impedances result in energy losses? (e) What is the expression for the equivalent series impedance of the transformer in terms of symbols of the above diagram? (f) If the lamination thickness of the core material in the transformer were doubled, which impedances in the equivalent circuit would be affected? Explain.
  Ans. (e) To a good approximation, the shunt branch is moved to the extreme left, as in Fig. 2-12. The series impedance is then $(R_1 + a^2R_2) + j(X_1 + a^2X_2)$. (f) Doubling the lamination thickness doubles the eddy-current loss. Thus, the value of R_c would decrease.

2.41. Which transformer would you expect to be the heavier, a 25-Hz unit or a 60-Hz unit of identical voltampere rating and identical secondary voltage rating? Explain.
  Ans. From (2.9) it follows that 25-Hz transformer requires a greater flux as compared to a 60-Hz. This implies a greater core cross section for the 25-Hz transformer (for the same B_{max}). Hence the 25-Hz transformer is heavier.

2.42. A transformer is rated 1-kVA, 240-V/120-V, 60-Hz. Because of an emergency, this transformer has to be used on a 50-Hz system. If the flux density in the transformer core is to be kept the same as at 60 Hz and 240 V: (a) What voltage should be applied at 50 Hz? (b) What is the voltampere rating at 50 Hz?   *Ans.* (a) 200 V;   (b) 0.833 kVA

2.43. An *ideal* transformer is rated 2400-V/240-V. A certain load of 50 A, unity power factor, is to be connected to the low-voltage winding. This load must have exactly 200 V across it. With 2400 V applied to the high-voltage winding, what resistance must be added in series with the transformer, if located (a) in the low-voltage winding, (b) in the high-voltage winding?   *Ans.* (a) 0.8 Ω;   (b) 80 Ω

2.44. From Problem 2.18, determine the power factor at which the voltage regulation of the transformer at full-load is zero.   *Ans.* 0.835 leading

2.45. A 50-kVA, 2300-V/230-V, 60-Hz transformer takes 200 W and 0.30 A at no-load when 2300 V is applied to the high-voltage side. The primary resistance is 3.5 Ω. Neglecting the leakage reactance drop, determine (a) the no-load power factor, (b) the primary induced voltage, (c) the magnetizing current, (d) the core-loss current component.   *Ans.* (a) 0.29;   (b) ≈2300 V;   (c) 0.286 A;   (d) 0.088 A

Chapter 3

Electromechanical Systems

3.1 ELECTROMECHANICAL ENERGY CONVERSION

An *electromechanical energy converter* transforms electrical energy into mechanical form, and vice versa. These devices are either *gross-motion devices*, such as electric motors and generators, or *incremental-motion devices*, such as electromechanical transducers. In this chpater, we shall consider only the latter type. Some examples of incremental-motion transducers are: microphones, loudspeakers, electromagnetic relays, transducers, etc.

The two basic magnetic-field effects resulting in the production of forces are (1) alignment of flux lines, and (2) interaction between magnetic fields and current-carrying conductors. Although these forces are mechanical, operating on material bodies that have no overall electric charge, they are ultimately of electrical (electronic) origin. Hence we refer to them as "electrical forces" and use for them the symbol F_e. Examples of "alignment" are shown in Fig. 3-1. In Fig. 3-1(a), the force on the ferromagnetic pieces causes them to align with the flux lines, thus shortening the magnetic flux path and reducing the reluctance. Figure 3-1(b) shows a simplified form of a reluctance motor, in which the electrical force tends to align the axis of the rotor with that of the stator. Figure 3-1(c) shows the alignment of two current-carrying coils. A few examples of

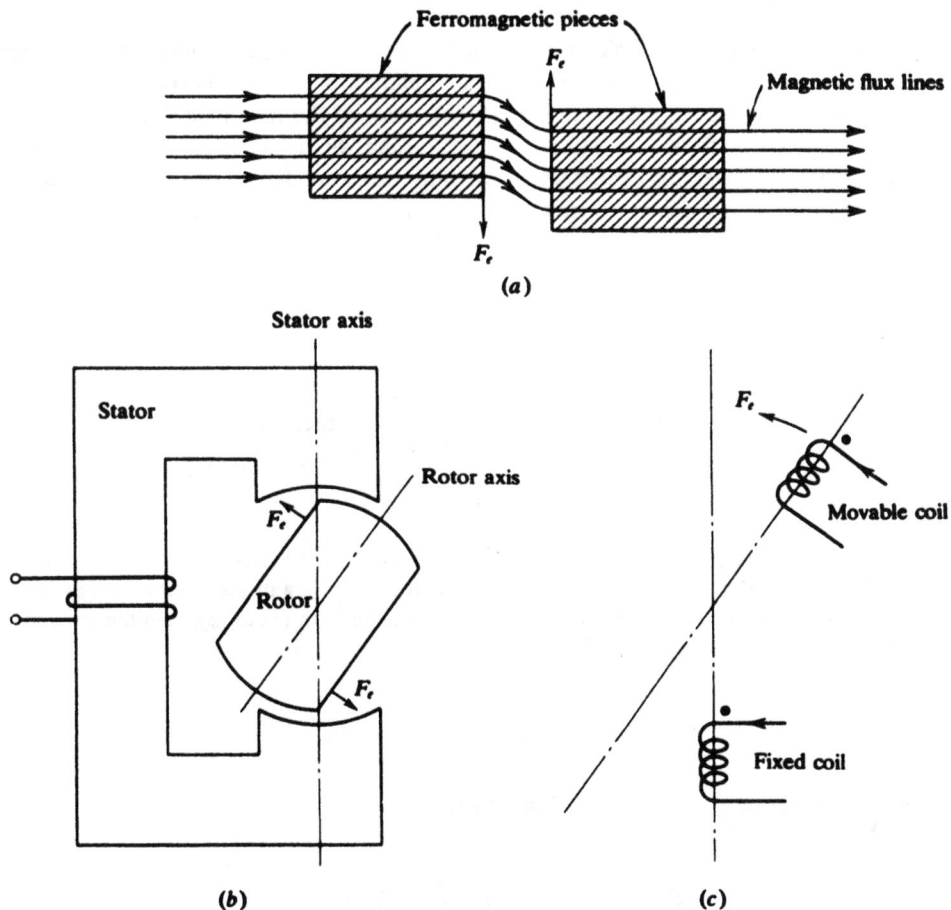

Fig. 3-1

"interaction" are shown in Fig. 3-2; here current-carrying conductors experience a force when placed in magnetic fields. For instance, in Fig. 3-2(b), the electrical force is produced by the interaction between the flux lines and coil current, resulting in a torque on the moving coil. This mechanism forms the basis of a variety of electrical measuring instruments. Almost all industrial dc motors work on the "interaction" principle.

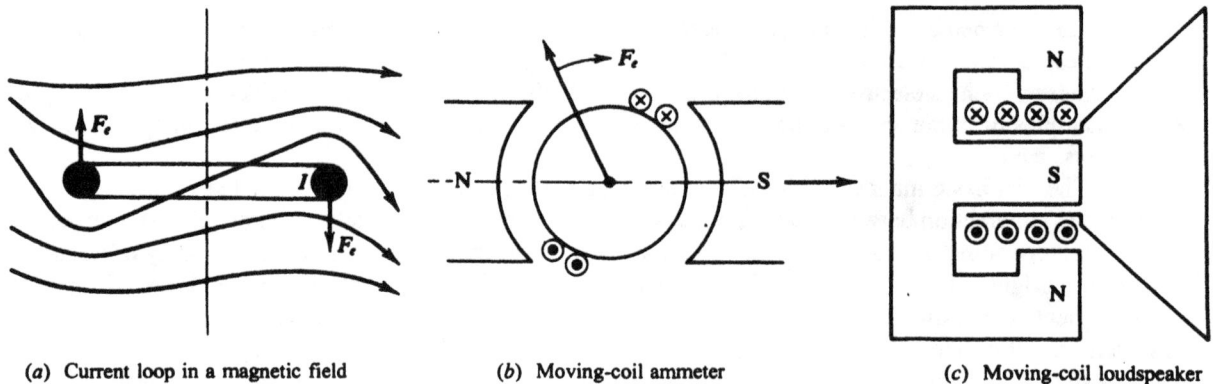

(a) Current loop in a magnetic field   (b) Moving-coil ammeter   (c) Moving-coil loudspeaker

Fig. 3-2

Quantitative evaluation of the electrical force will be considered presently; here we simply point out that the force is always in a direction such that the net magnetic reluctance is reduced or the energy stored in the magnetic field is minimized. Thus, in Fig. 3-1(b), most of the magnetic-field energy is stored in the airgap separating the rotor from the stator. This airgap field is often called the *coupling field*. Electromechanical energy conversion occurs when coupling fields are distributed in such a way that the stored magnetic energy changes with mechanical motion. From the viewpoint of energy conservation, we may say that in a lossless (or conservative) system

net energy input = increase in stored energy

or

$$\begin{pmatrix} \text{electrical} \\ \text{energy input} \end{pmatrix} = \begin{pmatrix} \text{mechanical} \\ \text{work done} \\ \text{by system} \end{pmatrix} + \begin{pmatrix} \text{increase in} \\ \text{stored energy} \end{pmatrix} \qquad (3.1)$$

Example 3.1 Consider the special case of an electromagnet attracting an iron mass, Fig. 3-3, where (1) and (2) indicate respectively the initial and final positions of the iron mass, which undergoes a displacement $-dx$ (against the positive x-direction). If the coil current stays constant at $i = I_0$ during the motion from (1) to (2), then the input electrical energy, dW_e, from the current source is given by Faraday's law, (2.1), as

$$dW_e = I_0 e\,dt = I_0(\lambda_2 - \lambda_1) \qquad (3.2)$$

The increase in stored magnetic energy, dW_m, is, from (1.22),

$$dW_m = \frac{1}{2}(L_2 - L_1)I_0^2 = \frac{1}{2}(\lambda_2 - \lambda_1)I_0 \qquad (3.3)$$

where we have assumed a linear magnetic circuit, $L = \lambda/i$. By (3.1),

$$dW_e = (-F_e)(-dx) + dW_m \qquad (3.4)$$

where F_e is the electrical force. Thus, from (3.2), (3.3), and (3.4),

$$F_e dx = \frac{1}{2} (\lambda_2 - \lambda_1) I_0 = dW_m \qquad (3.5)$$

If, on the other hand, the flux linkage stays constant at $\lambda = \lambda_0$ during the motion, we have instead of (3.2) and (3.3):

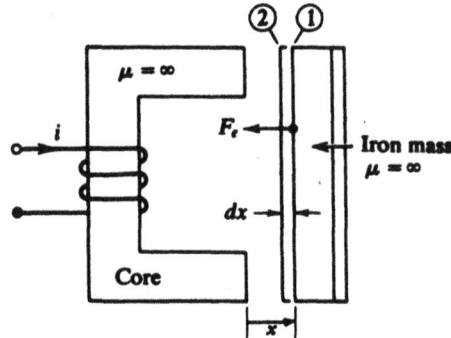

Fig. 3-3

$$dW_e = 0 \qquad (3.6)$$

$$dW_m = \frac{1}{2} \lambda_0 (i_2 - i_1) \qquad (3.7)$$

which, together with (3.4), yield

$$F_e dx = \frac{1}{2} \lambda_0 (i_2 - i_1) = -dW_m \qquad (3.8)$$

3.2 FORCE AND TORQUE EQUATIONS

We may rewrite (3.5) and (3.8) as

for current excitation: $\qquad F_e = \dfrac{\partial W_m(i, x)}{\partial x} \qquad (3.9)$

for voltage excitation: $\qquad F_e = - \dfrac{\partial W_m(\lambda, x)}{\partial x} \qquad (3.10)$

These are the two forms of the force equation, giving the value of the mechanical force of electrical origin. For rotary-motion systems, the analogous expressions for torque are

for current excitation: $\qquad T_e = \dfrac{\partial W_m(i, \theta)}{\partial \theta} \qquad (3.11)$

$$\text{for voltage excitation:} \quad T_e = -\frac{\partial W_m(\lambda, \theta)}{\partial \theta} \tag{3.12}$$

As is proved in Problem 3.42, (3.9) and (3.10)—or (3.11) and (3.12)—may be used interchangeably for a linear magnetic circuit.

3.3 ELECTROMECHANICAL DYNAMICS

The behavior of an electromechanical system is governed by the electrical and mechanical equations of motion. These two equations, which in general are coupled, are conveniently stated in the form of a *voltage-* (or *current-*) *balance equation* and a *force-* (or *torque-*) *balance equation*, wherein the "applied forces" are equated to the "restoring forces." (The restoring forces include the "inertial force" or "acceleration" of the system, so that the "balance" equations are actually equations of motion.) In the electrical equation, electrical forces of mechanical origin, such as induced emf as given by Faraday's law, play the part of applied forces. In the mechanical equation, mechanical forces of electrical origin, e.g., the force F_e given by (3.9) or (3.10), act as the applied force.

Example 3.2 An electromagnetic relay may be modeled by the lumped-parameter system shown in Fig. 3-4. There is no externally applied mechanical force. We formulate the dynamical equations of motion in the following steps.

1. *Assumptions.* We neglect saturation of the magnetic circuit, which is assumed infinitely permeable, and ignore leakage and fringing fluxes. Also, we assume the friction force to be directly proportional to velocity, and the spring force to be directly proportional to the elongation.

2. *Parameters.* The mechanical parameters are mass, M; friction coefficient, b; and spring stiffness, k. The parameters for the electrical circuit are resistance, R; and inductance, L, which may be expressed in terms of the dimensions shown in Fig. 3-4 as

$$L(x) = \frac{\mu_0 a N^2}{l_1 - x} \tag{3.13}$$

3. *Equations of motion.* We can now write the balance equations between the different "forces" acting on the system:

$$\text{electrical:} \quad Ri + \frac{d}{dt}(Li) = v \tag{3.14}$$

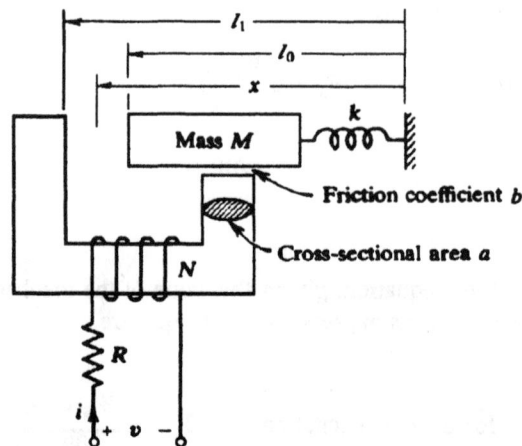

Fig. 3-4

where the terms on the left-hand side denote the restoring forces (voltage drops); and

mechanical: $M\ddot{x} + b\dot{x} + k(x - l_0) = F_e = \frac{1}{2} i^2 \frac{\partial L}{\partial x}$ (3.15)

where the left-hand side is the sum of the restoring forces and where F_e, considered as an external force, is given by (3.9) and (1.22).

Equations (3.14) and (3.15) yield the electromechanical dynamics of the system. However, these equations are nonlinear, and analytical solutions are not forthcoming. For small-signal, incremental motion, useful information about the system can be obtained by solving the corresponding linearized equations. Small-signal linearization is done about a steady-state operating point. In the present example, let (V_0, I_0, X_0) denote the steady-state, stable equilibrium point, such that

$$v(t) = V_0 + v_1(t)$$

$$i(t) = I_0 + i_1(t)$$

$$x(t) = X_0 + x_1(t)$$

where (v, i, x) are the original variables and (v_1, i_1, x_1) are small perturbations about (V_0, I_0, X_0). The smallness is measured by the fact that product-type terms such as i_1^2, i_1x_1, and so forth, are negligible in comparison to I_0^2, I_0X_0, and so forth. We substitute

$$x = X_0 + x_1$$

$$L_0 \equiv \frac{\mu_0 a N^2}{l_1 - X_0}$$

in (3.13) to obtain

$$L = L_0 \left(1 - \frac{x_1}{l_1 - X_0}\right)^{-1} = L_0 \left[1 + \frac{x_1}{l_1 - X_0} + \left(\frac{x_1}{l_1 - X_0}\right)^2 + \left(\frac{x_1}{l_1 - X_0}\right)^3 + \cdots\right]$$ (3.16)

The series converges because (by assumption) $x(t) < l_1$, i.e., $x_1 < l_1 - X_0$. Also, from (3.16),

$$\frac{\partial L}{\partial x} = \frac{\partial L}{\partial x_1} = \frac{L_0}{l_1 - X_0} \left[1 + \frac{2x_1}{l_1 - X_0} + 3\left(\frac{x_1}{l_1 - X_0}\right)^2 + \cdots\right]$$ (3.17)

The linearized forms of (3.16) and (3.17) would then be

$$L \approx L_0 \left(1 + \frac{x_1}{l_1 - X_0}\right)$$

$$\frac{\partial L}{\partial x} \approx \frac{L_0}{l_1 - X_0} \left(1 + \frac{2x_1}{l_1 - X_0}\right)$$ (3.18)

The flux linkage $\lambda = Li$ then becomes

$$Li \approx L_0 \left(1 + \frac{x_1}{l_1 - X_0}\right)(I_0 + i_1) \approx L_0 I_0 + L_0 i_1 + \frac{L_0 I_0}{l_1 - X_0} x_1$$ (3.19)

in which we have neglected the product of the small quantities $x_1/(l_1 - X_0)$ and i_1/I_0. Moreover,

$$\frac{d}{dt}(Li) \approx L_0 \frac{di_1}{dt} + \frac{L_0 I_0}{l_1 - X_0} \dot{x}_1 \tag{3.20}$$

Substituting (3.20), $Ri = RI_0 + Ri_1$, and $v = V_0 + v_1$ in (3.14), and recognizing that $RI_0 = V_0$ is the steady-state, or dc, operating point, we obtain as the linearized electrical equation of motion:

$$\text{linearized electrical:} \quad L_0 \frac{di_1}{dt} + Ri_1 + \frac{L_0 I_0}{l_1 - X_0} \dot{x}_1 = v_1 \tag{3.21}$$

Next, we consider the right-hand side of (3.15), in which we substitute (3.18) and $i^2 \approx I_0^2 + 2I_0 i_1$ to obtain

$$\frac{1}{2} i^2 \frac{\partial L}{\partial x} \approx \frac{1}{2}(I_0^2 + 2I_0 i_1) \frac{L_0}{l_1 - X_0}\left(1 + \frac{2x_1}{l_1 - X_0}\right)$$

$$\approx \frac{L_0 I_0^2}{2(l_1 - X_0)} + \frac{L_0 I_0}{l_1 - X_0} i_1 + \frac{L_0 I_0^2}{(l_1 - X_0)^2} x_1$$

in which the same sort of term was neglected as was dropped from (3.19). Equation (3.15) now becomes

$$M\ddot{x}_1 + b\dot{x}_1 + kx_1 + k(X_0 - l_0) = \frac{L_0 I_0^2}{2(l_1 - X_0)} + \frac{L_0 I_0}{l_1 - X_0} i_1 + \frac{L_0 I_0^2}{(l_1 - X_0)^2} x_1 \tag{3.22}$$

The steady-state mechanical equilibrium is given by

$$k(X_0 - l_0) = \frac{L_0 I_0^2}{2(l_1 - X_0)}$$

and so the remaining terms of (3.22) provide the linearized mechanical equation of motion:

$$\text{linearized mechanical:} \quad M\ddot{x}_1 + b\dot{x}_1 + \left[k - \frac{L_0 I_0^2}{(l_1 - X_0)^2}\right]x_1 = \frac{L_0 I_0}{l_1 - X_0} i_1 \tag{3.23}$$

The electromechanical dynamics of the system may now be ascertained by solving simultaneously (3.21) and (3.23), e.g., by use of the Laplace transform (see Problems 3.9 and 3.38). As an alternative, the original equations of motion, (3.14) and (3.15), may be solved by numerical integration (see Problem 3.14).

3.4 ELECTROMECHANICAL ANALOGIES

Mechanical systems can be represented by electrical circuits, or vice versa, via either of the analogies presented in Tables 3-1 and 3-2.

Table 3-1. Force-voltage analogy

Force, F	Voltage, v
Velocity, x	Current, i
Damping, b	Resistance, R
Mass, M	Inductance, L
Spring constant, k	Elastance = reciprocal of capacitance, $1/C$

Table 3-2. Force-current analogy

Force, F	Current, i
Velocity, x	Voltage, v
Damping, b	Conductance, G
Mass, M	Capacitance, C
Spring constant, k	Reciprocal of inductance, $1/L$

Figure 3-5 shows the two electrical analogs of a particular mechanical system.

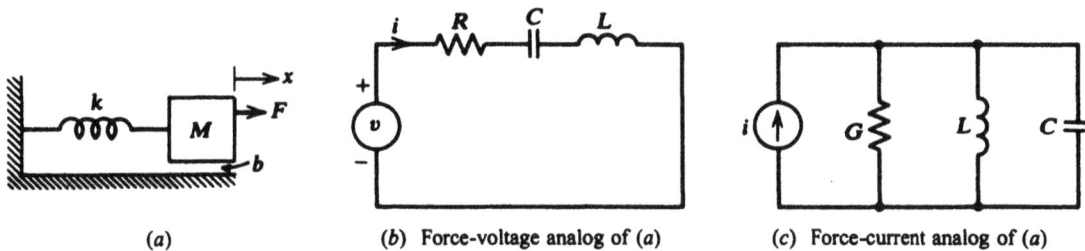

(a)   (b) Force-voltage analog of (a)   (c) Force-current analog of (a)

Fig. 3-5

Solved Problems

3.1. A solenoid of cylindrical geometry is shown in Fig. 3-6. (a) If the exciting coil carries a dc steady current I, derive an expression for the force on the plunger. (b) For the numerical values $I = 10$ A, $N = 500$ turns, $g = 5$ mm, $a = 20$ mm, $b = 2$ mm, and $l = 40$ mm, what is the magnitude of the force? Assume $\mu_{core} = \infty$ and neglect leakage.

For the magnetic circuit, the reluctance is

$$\Re = \frac{g}{\mu_0 \pi c^2} + \frac{b}{\mu_0 2\pi al} \quad \text{where} \quad c = a - \frac{b}{2}$$

The inductance L is then given by (1.20) as

$$L = \frac{N^2}{\Re} = \frac{2\pi\mu_0 alc^2 N^2}{2alg + bc^2} \equiv \frac{k_1}{k_2 g + k_3}$$

where $k_1 \equiv 2\pi\mu_0 alc^2 N^2$, $k_2 \equiv 2al$, and $k_3 \equiv bc^2$.

(a) Expressing the force as in (3.15),

$$F_e = \frac{1}{2} I^2 \frac{\partial L}{\partial g} = - \frac{I^2 k_1 k_2}{2(k_2 g + k_3)^2}$$

where the minus sign indicates that the force tends to decrease the airgap.

(b) Substituting the numerical values in the force expression of (a) yields 600 N as the magnitude of F_e.

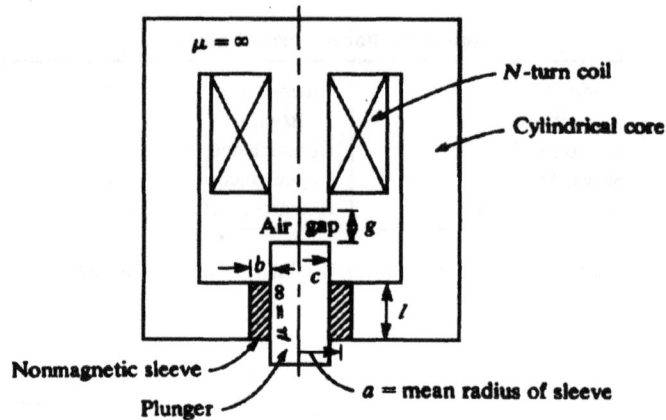

Fig. 3-6

3.2. (a) If the solenoid of Problem 3.1(a) instead carries an alternating current of 10 A (rms) at 60 Hz, what is the instantaneous force? (b) What is the average force, if N, g, a, b, and l have the same numerical values as in Problem 3.1(b)?

(a)   The instantaneous force is given by

$$F_e = -\frac{10\sqrt{2} \cos 120\pi t)^2 k_1 k_2}{2(k_2 g + k_3)^2} = -\frac{100 k_1 k_2}{(k_2 g + k_3)^2} \cos^2 120\pi t \quad (N)$$

(b)   Because the $\cos^2$ has average value 1/2, the average force is the same as the force due to 10 A dc, namely, 600 N.

3.3. Figure 3-7 shows a solenoid where the core cross section is square. (a) For a coil current of I (dc), derive an expression for the force on the plunger. (b) Given $I = 10$ A, $N = 500$ turns, $g = 5$ mm, $a = 20$ mm, and $b = 2$ mm, calculate the magnitude of the force.

(a)   From the equivalent electrical circuit, Fig. 1-16(c),

$$\text{reluctance:} \quad \mathfrak{R} = \frac{b + g}{2\mu_0 a^2}$$

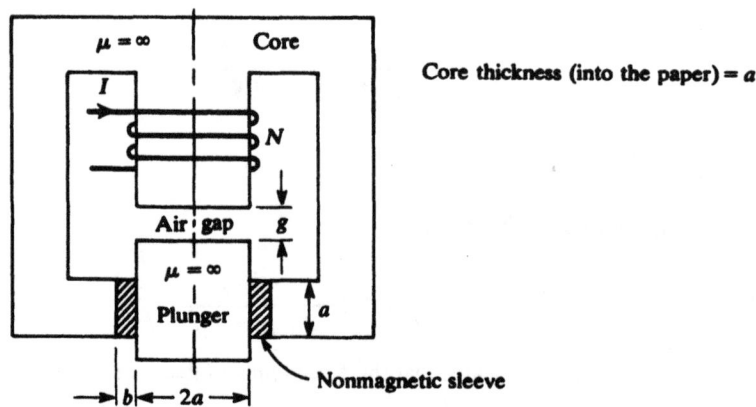

Fig. 3-7

Then, as in Problem 3.1(a),

$$\text{inductance:} \quad L = \frac{2\mu_0 a^2 N^2}{b + g}$$

$$\text{electrical force:} \quad F_e = \frac{1}{2} I^2 \frac{\partial L}{\partial g} = -\frac{\mu_0 a^2 N^2 I^2}{(b + g)^2}$$

(b)   256.4 N.

3.4. For a voltage-excited system, show that the electrical force can be expressed as

$$F_e = -\frac{1}{2} \phi^2 \frac{\partial \mathfrak{R}}{\partial x}$$

where ϕ is the core flux and $\mathfrak{R}$ is the net reluctance of the magnetic circuit.

We have

$$W_m = \frac{1}{2} L i^2 = \frac{1}{2} \frac{N\phi}{i} i^2 = \frac{1}{2} N\phi \left(\frac{\mathfrak{R}\phi}{N}\right) = \frac{1}{2} \mathfrak{R}\phi^2$$

and so, by (*3.10*), in which constant λ implies constant ϕ,

$$F_e = -\frac{\partial W_m}{\partial x} = -\frac{1}{2} \phi^2 \frac{\partial \mathfrak{R}}{\partial x}$$

3.5. Consider the solenoid shown in Fig. 3-7. Let the coil have a resistance R and be excited by a voltage $v = V_m \sin \omega t$. For a displacement g_0 between the plunger and the coil (pole face), determine the steady-state (a) coil current and (b) electrical force.

(a)   The electrical equation of motion of the system has the form (*3.14*):

$$Ri + \frac{d}{dt}(Li) = v$$

in which, from Problem 3.3(a),

$$L = \frac{2\mu_0 a^2 N^2}{b + g_0} = \text{constant}$$

Thus, we have to deal with the familiar

$$L \frac{di}{dt} + Ri = V \sin \omega t$$

The desired steady-state solution is simply $I = V/Z$, or

$$i = \frac{V_m}{\sqrt{R^2 + (\omega L)^2}} \sin(\omega t - \psi) \quad \text{where} \quad \psi = \arctan \frac{\omega L}{R}$$

(b)   Because the magnetic circuit is linear, we may determine the electrical force by (*3.9*), just as though the system were current-excited.

$$F_e = \frac{1}{2} i^2 \frac{\partial L}{\partial g_0} = - \frac{\mu_0 a^2 N^2}{(b + g_0)^2} i^2$$

where $i(t)$ is as found in (a) above.

3.6.   In Problem 3.5, for the numerical values $N = 500$ turns, $g_0 = 5$ mm, $a = 20$ mm, $b = 2$ mm, $R = 20$ Ω, $V_m = 120\sqrt{2}$ V, $\omega = 120\pi$ rad/s, calculate (a) the steady-state coil current and (b) the average steady-state electrical force.

(a)   At $g_0 = 5 \times 10^{-3}$ m,

$$L = \frac{2\mu_0 a^2 N^2}{b + g_0} = \frac{2(4\pi \times 10^{-7})(400 \times 10^{-6})(25 \times 10^4)}{7 \times 10^{-3}} = 0.036 \text{ H}$$

$$\omega L = 13.56 \ \Omega$$

$$Z = \sqrt{R^2 + (\omega L)^2} = \sqrt{400 + 183.88} = 24.16 \ \Omega$$

$$\psi = \arctan \frac{\omega L}{R} \approx 34°$$

$$i = \frac{V_m}{Z} \sin(\omega t - \psi) = 7.02 \sin(377t - 34°) \quad \text{(A)}$$

(b)   From (a),

$$. \ (i^2)_{avg} = \frac{(7.02)^2}{2} = 24.64$$

Then, from Problem 3.5,

$$(F_e)_{avg} = - \frac{\mu_0 a^2 N^2}{(b + g_0)^2} (i^2)_{avg} = - \frac{L}{2(b + g_0)} (i^2)_{avg} = -63.4 \text{ N}$$

3.7.   A two-winding system has its inductances given by

$$L_{11} = \frac{k_1}{x} = L_{22} \qquad L_{12} = L_{21} = \frac{k_2}{x}$$

where k_1 and k_2 are constants. Neglecting the winding resistances, derive an expression for the electrical force (as a function of x) when both windings are connected to the same voltage source, $v = V_m \sin \omega t$.

Because $L_{11} = L_{22}$ and $L_{12} = L_{21}$, we have $i_1 = i_2 \equiv i$. Hence the stored magnetic energy is

$$W_m = \frac{1}{2} L_{11} i_1^2 + \frac{1}{2} L_{22} i_2^2 + L_{12} i_1 i_2 = (L_{11} + L_{12}) i^2$$

and the electrical force is

$$F_e = \frac{\partial W_m(i, x)}{\partial x} = i^2 \frac{\partial}{\partial x} (L_{11} + L_{12}) = - \frac{(k_1 + k_2) i^2}{x^2} \qquad (1)$$

The current i is related to the voltage $v_1 = v_2 = v$ through

$$v = \frac{d}{dt}\left[(L_{11} + L_{12})i\right]$$

or

$$i = \frac{1}{L_{11} + L_{12}}\int v\, dt = \frac{x}{k_1 + k_2}\left(-\frac{V_m}{\omega}\cos \omega t\right) \tag{2}$$

Then (1) and (2) give

$$F_e = -\frac{V_m^2 \cos^2 \omega t}{(k_1 + k_2)\omega^2}$$

It is seen that F_e is apparently independent of x, a result which arises from having ignored the leakage flux.

3.8. Two mutually coupled coils are shown in Fig. 3-8. The inductances of the coils are: $L_{11} = A$, $L_{22} = B$, and $L_{12} = L_{21} = C \cos \theta$. Find the electrical torque for (a) $i_1 = I_0$, $i_2 = 0$; (b) $i_1 = i_2 = I_0$; (c) $i_1 = I_m \sin \omega t$, $i_2 = I_0$; (d) $i_1 = i_2 = I_m \sin \omega t$; and (e) coil 1 short-circuited and $i_2 = I_0$.

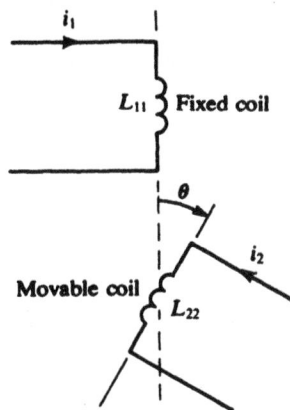

Fig. 3-8

(a)
$$W_m = \frac{1}{2} L_{11}I_0^2 \qquad T_e = \frac{\partial W_m}{\partial \theta} = 0$$

(b)
$$W_m = \frac{1}{2}(A + B)I_0^2 + CI_0^2 \cos \theta \qquad T_e = -CI_0^2 \sin \theta$$

(c)
$$W_m = \frac{1}{2} AI_m^2 \sin^2 \omega t + \frac{1}{2} BI_0^2 + CI_0 I_m \sin \omega t \cos \theta$$

$$T_e = -CI_0 I_m \sin \omega t \sin \theta$$

(d)
$$T_e = -CI_m^2 \sin^2 \omega t \sin \theta$$

(e) For coil 1:

$$\frac{d}{dt}(L_{11}i_1 + L_{12}i_2) = 0 \qquad or \qquad L_{11}i_1 + L_{12}i_2 = k = \text{constant}$$

Therefore, for given i_2 and L_{11},

$$i_1 = \frac{k - L_{12}I_0}{A}$$

and

$$W_m = \frac{A}{2}\left(\frac{k - L_{12}I_0}{A}\right)^2 + \frac{B}{2}I_0^2 + L_{12}I_0\left(\frac{k - L_{12}I_0}{A}\right)$$

$$= \frac{k^2}{2A} - \frac{L_{12}^2I_0^2}{2A} + \frac{B}{2}I_0^2 = \frac{k^2}{2A} + \frac{B}{2}I_0^2 - \frac{I_0^2}{2A}C^2\cos^2\theta$$

Hence,

$$T_e = \frac{\partial W_m}{\partial \theta} = \frac{I_0^2}{A}C^2\cos\theta\sin\theta$$

3.9. Figure 3-9 shows an angular-motion electromechanical system. The driving blade is made of iron and can move in the airgap, as shown. The blade is so designed that the inductance of the driving coil varies linearly with angular displacement; that is, $L = A + B\theta$, where A and B are constants. Other system parameters are: $R \equiv$ coil resistance, $J \equiv$ moment of inertia of the rotating parts, $b \equiv$ coefficient of friction between torsion rod and bearing, and $k \equiv$ torsional stiffness of the torsion rod. For a voltage input v, (a) write the equations of motion. (b) If these equations are nonlinear, identify the nonlinear terms and linearize about the steady-state operating point. (c) Taking angular displacement as the output and coil current as the input, obtain the *transfer function* of the system.

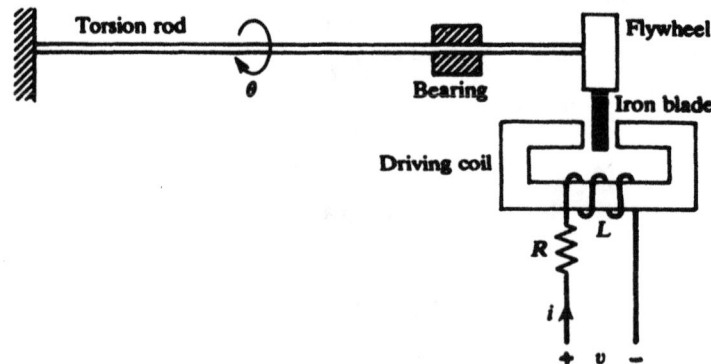

Fig. 3-9

(a)   The electrical equation is

$$(A + B\theta)\frac{di}{dt} + Bi\dot\theta + Ri = v \tag{1}$$

and the mechanical equation is

$$J\ddot\theta + b\dot\theta + k\theta = T_e = \frac{1}{2}Bi^2 \tag{2}$$

(b)   Nonlinear terms are

$$B\theta\frac{di}{dt} \quad\text{and}\quad Bi\dot\theta$$

in (1) and $1/2\, Bi^2$ in (2). To linearize, let

$$\theta = \Theta_0 + \theta_1$$

$$v = V_0 + v_i \qquad (3)$$

$$i = I_0 + i_1$$

where (Θ_0, V_0, I_0) is the steady-state equilibrium point and (θ_1, v_1, i_1) is the small time-dependent perturbation such that second-order terms may be neglected. Substitution of (3) into (1) and (2) yields the linearized equations

electrical: $\qquad (A + B\Theta_0)\dfrac{di_1}{dt} + BI_0\dot{\theta}_1 + Ri_1 = v_1 \qquad (4)$

mechanical: $\qquad J\ddot{\theta}_1 + b\dot{\theta}_1 + k\theta_1 = BI_0 i_1 \qquad (5)$

from which the relations determining the equilibrium point,

$$RI_0 = V_0 \qquad k\Theta_0 = \frac{1}{2}BI_0^2$$

have been subtracted.

(c)   The *transfer function*, $G(s)$, is defined as

$$G(s) \equiv \frac{\Theta_1(s)}{I_1(s)}$$

where $\Theta_1(s) \equiv$ Laplace transform of $I_1(t)$, and $I_1(s) \equiv$ Laplace transform of $i_1(t)$. Taking the Laplace transform of (5), on the assumption that $\theta_1(0) = \dot{\theta}_1(0) = 0$, gives

$$G(s) = \frac{BI_0}{Js^2 + bs + k}$$

3.10.   A capacitor microphone, shown in Fig. 3-10(a), may be modeled by the electromechanical system of Fig. 3-10(b). With v as the input, obtain the linearized dynamical equations of the system. Give the steady-state operating point.

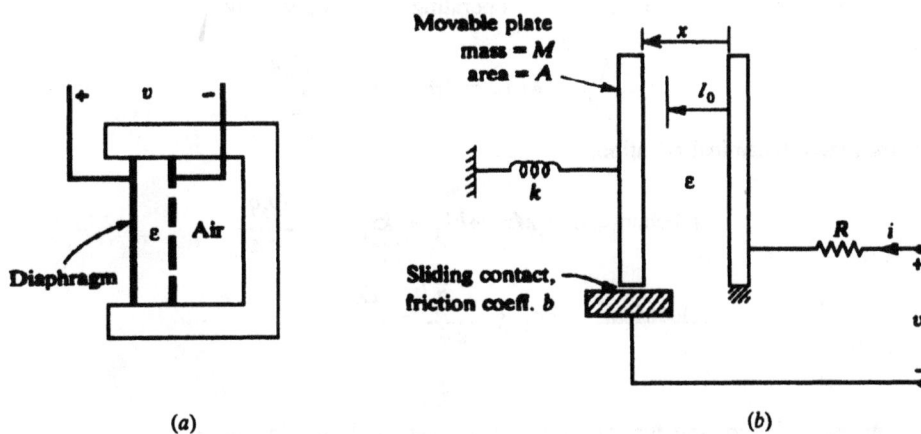

(a)   (b)

Fig. 3-10

For a charge q, the electrical energy stored in a capacitor can be expressed as

$$W_e = \frac{1}{2} \frac{q^2}{C}$$

where C is the capacitance. The electrical force between the plates is then given by

$$F_e = -\frac{\partial W_e}{\partial x} = -\frac{q^2}{2} \frac{\partial}{\partial x}\left(\frac{1}{C}\right)$$

where the minus sign indicates that the force tends to decrease the separation x of the plates. For the present case,

$$C = \frac{\varepsilon A}{x}$$

so that

$$F_e = -\frac{q^2}{2\varepsilon A}$$

The equations of motion are then:

mechanical: $\qquad M\ddot{x} + b\dot{x} + k(x - l_0) = -\dfrac{q^2}{2\varepsilon A}$ \hfill (1)

electrical: $\qquad R\dot{q} + \dfrac{qx}{\varepsilon A} = v$ \hfill (2)

These equations are nonlinear due to the presence of terms involving q^2 and qx.

Linearization is accomplished by assuming that a steady-state operating point (X_0, Q_0, V_0) exists such that

$$x = X_0 + x_1 \qquad q = Q_0 + q_1 \qquad v = V_0 + v_1$$

Substitution into (1) and (2), with neglect of higher-order terms, yields

$$M\ddot{x}_1 + b\dot{x}_1 + k(X_0 - l_0) + kx_1 = -\frac{Q_0^2}{2\varepsilon A} - \frac{Q_0 q_1}{\varepsilon A}$$

$$R\dot{q}_1 + \frac{Q_0 X_0}{\varepsilon A} + \frac{X_0 q_1}{\varepsilon A} + \frac{Q_0 x_1}{\varepsilon A} = V_0 + v_1$$

From these, it is seen that the steady-state operating point is given by

$$-\frac{Q_0^2}{2\varepsilon A} = k(X_0 - l_0) \qquad V_0 = \frac{Q_0 X_0}{\varepsilon A}$$

The linearized dynamical equations are thus

mechanical: $\qquad M\ddot{x}_1 + b\dot{x}_1 + kx_1 = -\dfrac{Q_0 q_1}{\varepsilon A}$

electrical: $\qquad R\dot{q}_1 + \dfrac{X_0 q_1}{\varepsilon A} + \dfrac{Q_0 x_1}{\varepsilon A} = v$

3.11   Draw an electrical analog for the mechanical system shown in Fig. 3-11.

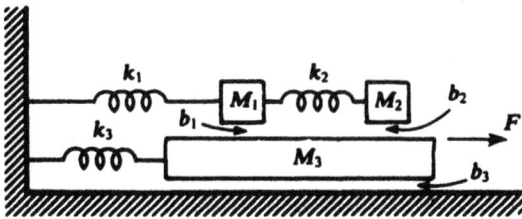

Fig. 3-11

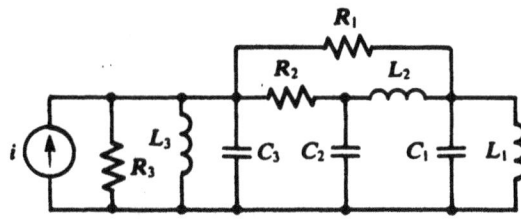

Fig. 3-12

An analog is given in Fig. 3-12, which is on the force-current analogy (see Table 3-2).

3.12   Draw a complete electrical equivalent circuit for the system shown in Fig. 3-9.

The circuit is shown in Fig. 3-13, where $L_1 \leftrightarrow A + B\Theta_0$, $L_2 \leftrightarrow (BI_0)^2/k$, $C_2 \leftrightarrow J/(BI_0)^2$, $G_2 \leftrightarrow b/(BI_0)^2$, and $v_2 \leftrightarrow BI_0\dot{\theta}_1$.

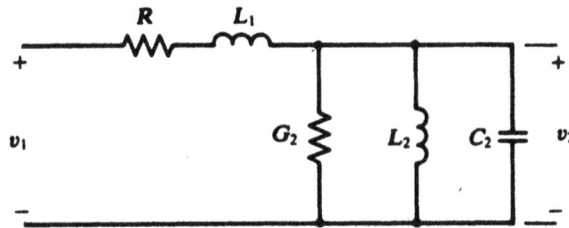

Fig. 3-13

3.13   The driving coil of the system shown in Fig. 3-4 has a negligible resistance and is excited by a voltage source $v = V_m \cos \omega t$. (a) Obtain an expression for $x(t)$ in the steady state, and (b) find the power supplied by the voltage source if the magnetic material of the core is ideal.

(a)   For a voltage-excited system (Problem 3.4),

$$F_e = -\frac{1}{2} \phi^2 \frac{\partial \mathfrak{R}}{\partial x}$$

In the present case

$$\mathfrak{R} = \frac{l_1 - x}{\mu_0 a} \quad \text{and} \quad \phi^2 = \left(\frac{V_m}{\omega N}\right)^2 \sin^2 \omega t$$

so that

$$F_e = \frac{1}{2\mu_0 a} \left(\frac{V_m}{\omega N}\right)^2 \sin^2 \omega t = A(1 - \cos 2\omega t) \quad \text{where} \quad A \equiv \frac{1}{4\mu_0 a} \left(\frac{V_m}{\omega N}\right)^2$$

Note that the force is in the positive x-direction. The mechanical equation of motion, (3.15), thus takes the form

$$M\ddot{x} + b\dot{x} + k(x - l_0) = A(1 - \cos 2\omega t)$$

The constant terms may be removed by shifting the origin of x; thus,

$$M\ddot{x} + b\dot{x} + kx = -A \cos 2\omega t \tag{1}$$

The steady-state solution of (1) may be written down from the force-voltage analogy (Table 3-1). The mechanical impedance, at a frequency 2ω, is expressed as

$$Z_m = \sqrt{b^2 + \left(2\omega M - \frac{k}{2\omega}\right)^2}$$

and the desired solution is

$$x(t) = -\frac{A}{Z_m}\cos(2\omega t - \psi) \quad \text{where} \quad \psi = \arctan\left[\frac{1}{b}\left(2\omega M - \frac{k}{2\omega}\right)\right]$$

(b)　　　　average power supplied = average power dissipated in friction = $b\dot{X}^2$ where $\dot{X}$ is the rms value of $\dot{x}$. From (a),

$$\dot{x} = \frac{2\omega A}{Z_m}\sin(2\omega t - \psi) \quad \text{and} \quad \dot{X} = \frac{\sqrt{2}\,\omega a}{Z_m}$$

Hence

$$P_{avg} = \frac{2b\omega^2 A^2}{Z_m^2}$$

3.14. For the system considered in Example 3.2, we have the following numerical values: $M = 0.01$ kg, $b = 0.1$ N$\cdot$s/m, $k = 100$ N/m, $l_0 = 20$ mm, $l_1 = 30$ mm, $N = 200$ turns, $a = 100$ mm^2, and $R = 1$ Ω. Express the original equations of motion as a set of state equations, and solve (numerically) for the current $i(t)$ for a 6-V step input.

Let the state variables be $y_1 = i$, $y_2 = x$, $y_3 = \dot{x}$. From (3.13),

$$L = \frac{\mu_0 a N^2}{l_1 - x} = \frac{(4\pi \times 10^{-7})(100 \times 10^{-6})(200)^2}{30 \times 10^{-3} - y_2} = \frac{16\pi \times 10^{-7}}{0.030 - y_2} \quad \text{(H)}$$

$$\frac{\partial L}{\partial x} = \frac{16\pi \times 10^{-7}}{(0.030 - y_2)^2} \quad \text{(H/m)}$$

$$\frac{dL}{dt} = \frac{16\pi \times 10^{-7}}{(0.030 - y_2)^2} y_3 \quad \text{(H/s)}$$

Using these and the numerical data in (3.14) and (3.15) yields the desired state equations:

$$\dot{y}_1 = \frac{0.030 - y_2}{16\pi \times 10^{-7}}\left[6 - y_1 - \frac{(16\pi \times 10^{-7})y_1 y_3}{(0.030 - y_2)^2}\right]$$

$$\dot{y}_2 = y_3$$

$$\dot{y}_3 = \frac{(8\pi \times 10^{-5})y_1^2}{(0.030 - y_2)^2} - 10^4(y_2 - 0.020) - 10y_3$$

together with the initial conditions (corresponding to equilibrium)

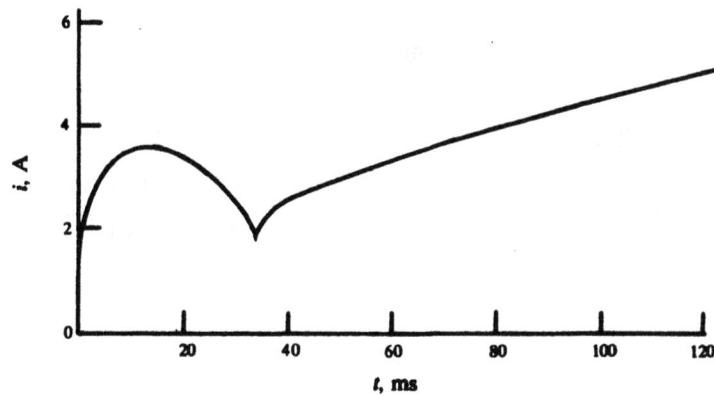

Fig. 3-14

$$y_1(0) = 0 \text{ A} \qquad y_2(0) = 0.020 \text{ m} \qquad y_3(0) = 0 \text{ m/s}$$

Numerical integration of the state equations can be performed on a digital computer; Fig. 3-14 shows the solution obtained for the current.

3.15. Up to this point, we have used the force equation (*3.9*), which was derived on the assumption that the magnetic circuit was linear. Derive the general force equation where the magnetic circuit is nonlinear, and show that (*3.9*) is a special case of the general equation.

Here, our starting point shall be (*3.4*), in which we substitute

$$dW_e = iv \, dt = i \, d\lambda$$

to obtain

$$F_e dx = -dW_m + i \, d\lambda \tag{3.24}$$

If we let i and x be the independent variables, the total differentials on the right of (*3.24*) become

$$d\lambda = \frac{\partial \lambda}{\partial i} \, di + \frac{\partial \lambda}{\partial x} \, dx \qquad dW_m = \frac{\partial W_m}{\partial i} \, di + \frac{\partial W_m}{\partial x} \, dx$$

giving

$$F_e dx = \left(-\frac{\partial W_m}{\partial x} + i \frac{\partial \lambda}{\partial x} \right) dx + \left(-\frac{\partial W_m}{\partial i} + i \frac{\partial \lambda}{\partial i} \right) di \tag{3.25}$$

Because di and dx are arbitrary, F_e must be independent of these changes. Thus, for F_e to be independent of di, its coefficient in (*3.25*) must be zero. Hence,

$$F_e = -\frac{\partial W_m(i, x)}{\partial x} + i \frac{\partial \lambda(i, x)}{\partial x} \tag{3.26}$$

which is the general force equation.

For a linear magnetic circuit, the energy stored in the inductor is $W_m = \frac{1}{2}i^2 L(x)$, whence

$$\frac{\partial W_m}{\partial x} = \frac{1}{2} i^2 \frac{\partial L}{\partial x} \tag{1}$$

where L is not a function of i. Moreover, since $\lambda = iL$,

$$i \frac{\partial \lambda}{\partial x} = i^2 \frac{\partial L}{\partial x} \qquad (2)$$

Substituting (*1*) and (*2*) in (*3.26*) yields

$$F_e = -\frac{1}{2} i^2 \frac{\partial L}{\partial x} + i^2 \frac{\partial L}{\partial x} = \frac{1}{2} i^2 \frac{\partial L}{\partial x} = \frac{\partial W_m(i, x)}{\partial x}$$

which is just (*3.9*).

3.16. Suppose the *i*-λ relationship for the electromechanical system of Fig. 3-3 to be

$$i = a\lambda^2 + b\lambda(x - c)^2$$

where *a*, *b*, *c* are constants. Derive a formula for the force on the iron mass at *x* = *g*.

Here we have a nonlinear magnetic circuit, with λ and *x* as the independent variables. Equation (*3.24*) becomes

$$F_e dx = -\left(\frac{\partial W_m}{\partial \lambda} d\lambda + \frac{\partial W_m}{\partial x} dx \right) + i \, d\lambda = -\frac{\partial W_m}{\partial x} dx + \left(i - \frac{\partial W_m}{\partial \lambda} \right) d\lambda$$

For F_e to be independent of *d*λ, we must have

$$\frac{\partial W_m}{\partial \lambda} = i \qquad \text{or} \qquad W_m = \int i \, d\lambda + w(x) \qquad (3.27)$$

and

$$F_e = -\frac{\partial W_m(\lambda, x)}{\partial x} \qquad (3.28)$$

[Note that (*3.28*) agrees with (*3.10*).] The unknown function *w*(*x*) in (*3.27*) may be equated to zero, as it represents magnetic energy at zero flux. Hence, for the given *i*-λ relationship,

$$W_m = \int \left[a\lambda^2 + b\lambda(x - c)^2 \right] d\lambda = \frac{a}{3} \lambda^3 + \frac{b}{2} \lambda^2 (x - c)^2$$

and

$$F_e = -\frac{\partial W_m}{\partial x} \bigg|_{x=g} = b\lambda^2(c - g)$$

3.17. For the magnetic circuit of a certain electromechanical system the flux linkage and current are related by

$$\lambda = \frac{2(i^{1/2} + i^{1/3})}{x + 1}$$

where *x* is a measure of displacement between fixed and moving members. Calculate the electrical force at *x* = 0 and *i* = 64 A dc.

Because the λ-*i* relationship is nonlinear, we must use the general force equation, (*3.26*). The vanishing of the coefficient of *di* in (*3.25*) implies that

$$\frac{\partial W_m}{\partial i} = i\,\frac{\partial \lambda}{\partial i}$$

whence

$$W_m = \int i\,\frac{\partial \lambda}{\partial i}\,di = \frac{2}{x+1}\int i\left(\frac{1}{2}\,i^{-1/2} + \frac{1}{3}\,i^{-2/3}\right)di = \frac{2}{x+1}\left(\frac{1}{3}\,i^{3/2} + \frac{1}{4}\,i^{4/3}\right)$$

Thus,

$$-\frac{\partial W_m}{\partial x} = \frac{2}{(x+1)^2}\left(\frac{1}{3}\,i^{3/2} + \frac{1}{4}\,i^{4/3}\right) \quad\text{and}\quad i\,\frac{\partial \lambda}{\partial x} = -\frac{2}{(x+1)^2}\left(i^{3/2} + i^{4/3}\right)$$

and (3.26) gives

$$F_e = -\frac{1}{(x+1)^2}\left(\frac{4}{3}\,i^{3/2} + \frac{3}{2}\,i^{4/3}\right)$$

At $x = 0$ and $i = 64$ A, $F_e = -1066.67$ N, tending to decrease x.

3.18. A permanent-magnet, moving-coil dc ammeter has moment of inertia J, torsional stiffness k, and friction coefficient b. (a) With the pointer deflection θ defined as the output, derive an expression for the transfer function of the meter. Let the meter torque constant be a (N · m/A). (b) Given the values $J = 3 \times 10^{-8}$ kg · m^2, $k = 2 \times 10^{-8}$ N · m/rad, and $b = 4 \times 10^{-8}$ N · m · s/rad, calculate the damping ratio and the natural frequency of undamped oscillations. (c) If $a = 2 \times 10^{-8}$ N · m/A and the other parameters are as in (b), obtain the unit step response of the meter.

(a)   If i is the input current, the mechanical equation of motion becomes

$$J\ddot{\theta} + b\dot{\theta} + k\theta = ai$$

from which the transfer function is obtained as

$$G(s) = \frac{\Theta(s)}{I(s)} = \frac{a}{Js^2 + bs + k} = \frac{a/J}{s^2 + 2\zeta\omega_n s + \omega_n^2}$$

where $\omega = \sqrt{k/J} \equiv$ frequency of undamped oscillations

$$\zeta = \frac{b}{2\sqrt{kJ}} \equiv \text{damping ratio}$$

(b)   From the above definitions, for the given numerical values:

$$\omega_n = \sqrt{\frac{2}{3}} = 0.816 \text{ rad/s} \qquad \zeta = \frac{4}{2\sqrt{2\times 3}} = 0.816$$

(c)   For the given numerical values,

$$G(s) = \frac{2}{3s^2 + 4s + 2}$$

and a unit current-step has Laplace transform $I(s) = 1/s$. Hence,

$$\Theta(s) = \frac{2}{s(3s^2 + 4s + 2)}$$

from which

$$\theta(t) = 1 - 1.73e^{-2t} \cos(\sqrt{2}\,t - 54.6°)$$

Supplementary Problems

3.19. An electromagnet, shown in Fig. 3-15, is required to exert a 500-N force on the iron at an airgap of 1 mm, while the exciting coil is carrying 25 A dc. The core cross section at the airgap is 600 mm² in area. Calculate the required number of turns of the exciting coil.   *Ans.* 65 turns

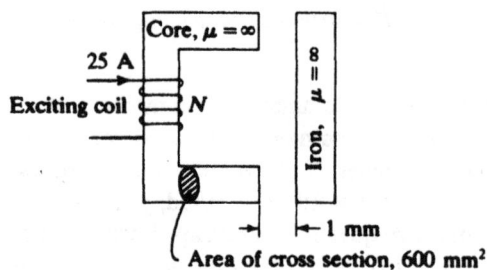

Fig. 3-15

3.20. (a) How many turns must the exciting coil of the electromagnet of Fig. 3-15 have in order to produce a 500-N (average) force if the coil is excited by a 60-Hz alternating current having a maximum value of 35.35 A? (b) Is the average force frequency-dependent?   *Ans.* (a) 65 turns;   (b) no

3.21. Figure 3-16 shows two mutually coupled coils, for which

$$L_{11} = L_{22} = 3 + \frac{2}{3x}\ (\text{mH}) \qquad L_{12} = L_{21} = \frac{1}{3x}\ (\text{mH})$$

where x is in meters. (a) If $i_1 = 5$ A dc and $i_2 = 0$, what is the mutual electrical force between the coils at $x = 0.01$ m? (b) If $i_1 = 5$ A dc and the second coil is open-circuited and moved in the positive x-direction at a constant speed of 20 m/s, determine the voltage across the second coil at $x = 0.01$ m. *Ans.* (a) 83.33 N;   (b) 333.3 V

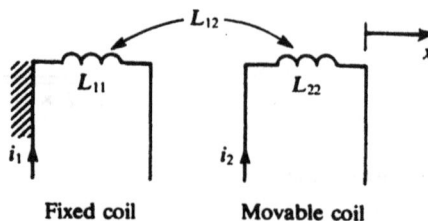

Fig. 3-16

3.22. For the two-coil system of Fig. 3-16, if $i_1 = 7.07 \sin 377t$ (A), $i_2 = 0$, and $x = 0.01$ m, determine (a) the instantaneous and (b) the time-average electrical force.   *Ans.* (a) 166.67 $\sin^2 377t$ (N); (b) 83.33 N

3.23. (a) The two coils of Fig. 3-16 are connected in series, with 5 A dc flowing in them. Determine the electrical force between the coils at $x = 0.01$ m. Does the force tend to increase or decrease x? (b) Next, the coils are connected in parallel across a 194-V, 60-Hz source. Compute the average electrical force at $x = 0.01$ m, neglecting the coil resistances.   *Ans.* (a) 250 N, decrease x;   (b) 250 N

3.24. A solenoid may be represented by an iron core inductor. The inductance as a function of the position of the plunger is $L(x) = (200 + 50\,x)$ mH. If a voltage $v = 100 \cos 100t$ (V) is applied across the inductor, calculate (a) the instantaneous force and (b) the average force exerted on the plunger at $x = 20$ cm.
Ans. (a) 0.567 $\sin^2 100t$ N;   (b) 0.283 N

3.25. The inductances of a 2-coil system, such as that shown in Fig. 3-16 are:

$$L_{11} = L_{22} = 4 \text{ mH} \qquad L_{12} = L_{21} = (1 - x^2) \text{ mH}$$

(a) The current in coil 1 is $i_1 = 5 \sin t$ (A). Under this condition at $x = 0.5$ and $x = 1$ m/s with coil 2 open-circuited, calculate the voltage across the terminals of coil 2. (b) Next, coil 2 is short-circuited and $i_1 = 5$ A dc. What is time-average force between the two coils?   *Ans.* (a) 4.25 $\cos t - 5 \sin t$ (V); (b) $50x (1 - x^2)\, 10^{-3}$ N

3.26. Refer to the solenoid of Fig. 3-7. A 100-turn coil is wound on one of the outer limbs of the solenoid. The N-turn coil (on the center limb) carries 5 A dc. Calculate the induced voltage in the 100-turn (outer limb) coil if the plunger moves at 0.5 m/s velocity.   *Ans.* 3.2 V

3.27. The device shown in Fig. 3-17(a) is a single-phase reluctance motor. The stator inductance varies sinusoidally with rotor position, as shown in Fig. 3-17(b). If the stator is supplied with a current $i = I_m \sin \omega t$ and the rotor rotates at ω_r (rad/s), (a) derive an expression for the instantaneous torque developed by the rotor. (b) Notice from (a) that the time-average torque (in the limit as the time interval becomes infinite) is zero, unless a certain condition is fulfilled. What is that condition? Express the nonzero average torque in terms of I_m, L_d, L_q, and the angle δ. (c) Given $L_d = 2L_q = 200$ mH and $I_m = 8$ A, what is the maximum value of the nonzero average torque?   *Ans.* (b) $\omega = \omega_r$; (c) 0.8 N $\cdot$ m

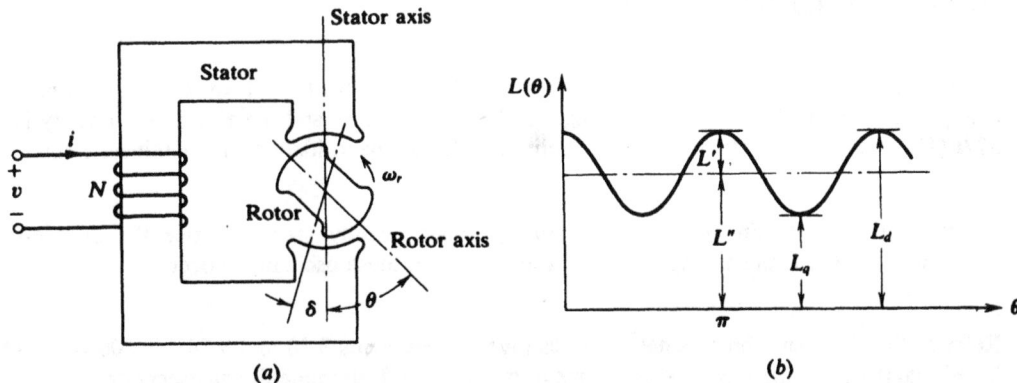

Fig. 3-17

3.28. The core of the electromagnet of Fig. 3-15 has a saturation characteristic that may be approximated as $\lambda = 2 \times 10^{-3} \sqrt{i}/x$, where all quantities are in SI units. Calculate the force developed by the magnet if the coil current is 5 A dc and $x = 10^{-3}$ m.   *Ans.* −15 N

3.29. Refer to Fig. 3-17. The motor runs at 3600 rpm and develops a maximum torque of 1 N · m while being fed by a constant-current source. The motor parameters are: $L_d = 2.5$ $L_q = 160$ mH, stator winding resistance $R = 0.2$ Ω, and bearing friction coefficient $b = 1.1 \times 10^{-4}$ N · m · s/rad. Neglecting the core losses, compute the total losses. *Ans.* 24 W

3.30. A cylindrical electromagnet is shown in Fig. 3-18. Given: $a = 2$ mm, $c = 40$ mm, $l = 40$ mm, $N = 500$ turns, and $R = 3.5$ Ω. A 110-V, 60-Hz source is applied across the coil. At $x = 5$ mm, determine (a) the maximum airgap flux density and (b) the average value of the electrical force. Assume $\mu_i \gg \mu_0$. *Ans.* (a) 0.65 T; (b) 106 N

3.31. From the data of Problem 3.30 calculate the steady-state (rms) value of (a) the coil current, (b) the input power, and (c) the power factor. *Ans.* (a) 4 A; (b) 56.56 W; (c) 0.13 lagging

3.32. In Fig. 3-18, let the spring be detached. The data are as in Problem 3.30. Initially the airgap is 5 mm; it is reduced to 2 mm while the coil is carrying a constant 4 A dc. Calculate (a) the energy supplied by the electrical source and (b) the mechanical work done in moving the mass. *Ans.* (a) 1.4 J; (b) 0.7 J

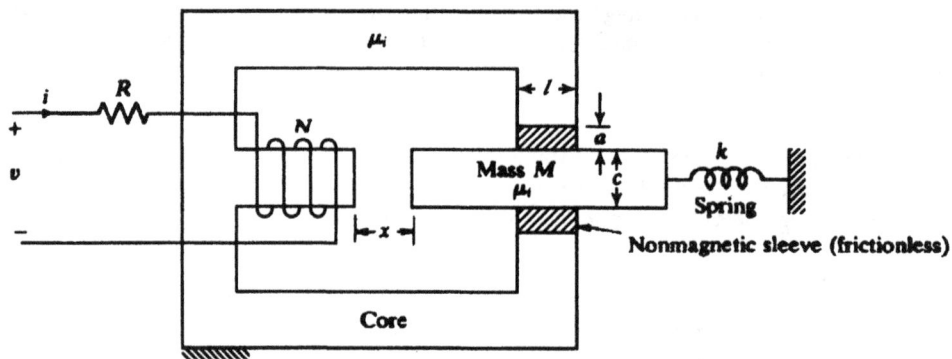

Fig. 3-18

3.33. The flux density in the 5-mm airgap of the electromagnet of Problem 3.30 is given as 0.65 sin 377t (T). Determine (a) the average value of the electrical force and (b) the rms value of the induced voltage of the coil. *Ans.* (a) 106 N; (b) 109 V

3.34. Write the dynamical equation of the system of Problem 3.30, if $M = 0.1$ kg, $k = 22.5$ kN/m, friction is negligible, the natural length of the spring (for $i = 0$) is 25 mm, and the airgap flux density is 0.65 sin 377t (T). Are these equations linear or nonlinear? If nonlinear, identify the nonlinear terms.

3.35. In Problem 3.34, obtain an expression for $x(t)$ under steady state. Notice the double-frequency oscillations. What is the average value of x about which these oscillations occur? *Ans.* 23.2 mm

3.36. Refer to Fig. 3-18 and the parametric values given in Problems 3.30 and 3.34. A 200-A direct current is fed into the coil. (a) Write the equations of motion. (b) Determine the quiescent operating point. (c) Linearize the equations about the quiescent operating point.
Ans. (b) $V_0 = 700$ V, $I_0 = 200$ A, $X_0 = 24.4$ mm

3.37. For the system of Problem 3.30, plot a set of curves of electrical force versus x for values of direct current ranging from 100 A to 700 A, at 200-A intervals. On the same graph superimpose the curve of spring force versus x. Hence, determine the quiescent operating points for each current, and investigate the stability of the system for each quiescent point thus determined.

3.38. In Problem 3.36, define x as the output and v as the input (instead of 200 A dc). Obtain the transfer function from the linearized equations of motion.

Ans.
$$G(s) = \frac{\beta}{(Ms^2 + bs + k)(Ls_1 + R) + \beta^2 s}$$

where

$$\beta \equiv \frac{4I_0}{\pi\mu_0 c^2}$$

$$L_1 \equiv \frac{1}{\pi\mu_0 c}\left(\frac{4X_0}{c} + \frac{a}{l}\right)$$

and (X_0, I_0) is the quiescent operating point.

3.39. Represent the system of Fig. 3-18 by a completely electrical equivalent circuit.
Ans. See Fig. 3-19, which is based on the force-current analogy.

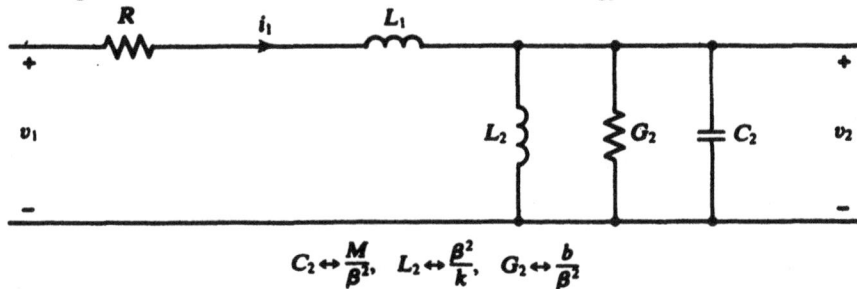

$$C_2 \leftrightarrow \frac{M}{\beta^2}, \quad L_2 \leftrightarrow \frac{\beta^2}{k}, \quad G_2 \leftrightarrow \frac{b}{\beta^2}$$

Fig. 3-19

3.40. Obtain electrical analogs for the mechanical systems shown in Fig. 3-20. *Ans.* See Fig. 3-21.

3.41. Linearize the following equations for a small perturbation (x_1, i_1, v_1) about the quiescent operating point (X_0, I_0, V_0):

(a) $\dfrac{d}{dt}\left(\dfrac{i}{x}\right) + 3i + 4ix = v$

(b) $\dfrac{d^2}{dt^2}\left(\dfrac{1}{x}\right) + 2x^2 + i^2 = 0$

(c) $\dfrac{di}{dt} + 4i + 2i\dfrac{dx}{dt} = v$

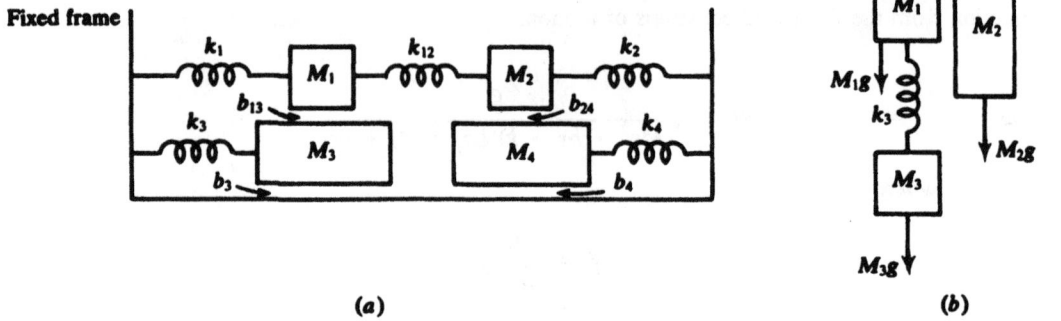

Fig. 3-20

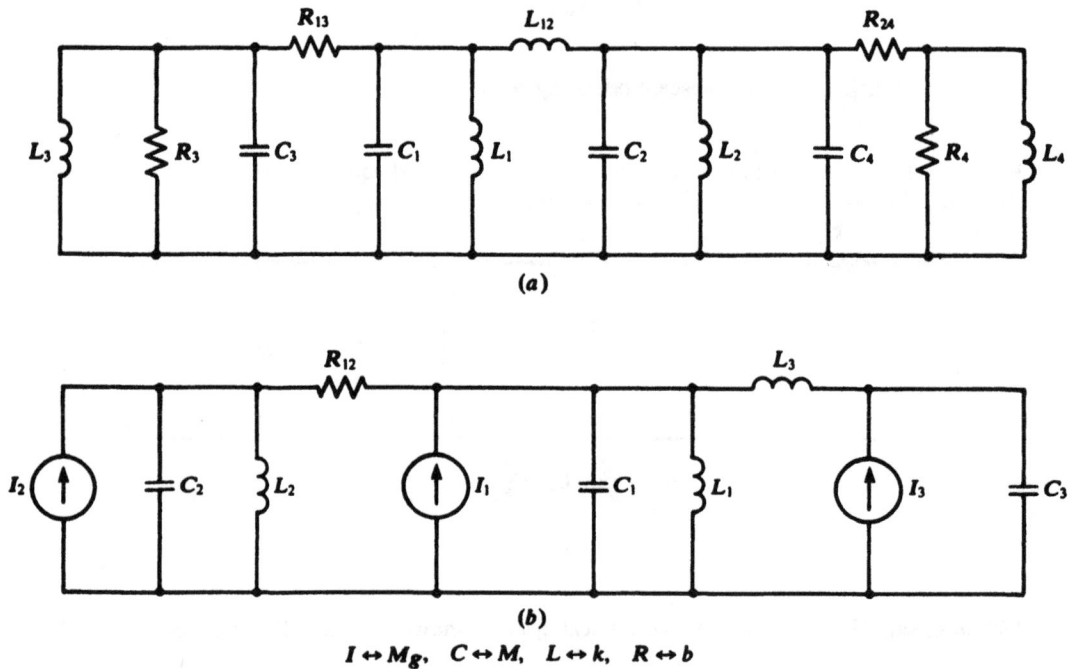

$$I \leftrightarrow Mg, \quad C \leftrightarrow M, \quad L \leftrightarrow k, \quad R \leftrightarrow b$$

Fig. 3-21

3.42. For a linear magnetic circuit,

$$W_m = \frac{i^2 L(x)}{2} = \frac{\lambda^2}{2L(x)}$$

Show that

$$\frac{\partial W_m(i, x)}{\partial x} = -\frac{\partial W_m(\lambda, x)}{\partial x}$$

Chapter 4

DC Machines

4.1 OPERATING PRINCIPLES

As indicated in Chapter 3, most electric machines operate on the basis of interaction between current-carrying conductors and electromagnetic fields. In particular, generator action is based on Faraday's law of electromagnetic induction, (2.1), which implies that a voltage (emf) is induced in a conductor moving in a region having flux lines at right angles to the conductor. That is, if a straight conductor of length l moves at velocity $\mathbf{u}$ (normal to its length) through a uniform magnetic field $\mathbf{B}$, the conductor itself always at right angles to $\mathbf{B}$, then only the velocity component $u_\perp$ orthogonal to $\mathbf{B}$ is effective in inducing the voltage e. In fact, the *Blu-rule* states:

$$e = Blu_\perp \qquad (4.1)$$

It follows that the voltage e induced in an N-turn rectangular coil, of axial length l and radius r, rotating at a constant angular velocity ω in a uniform magnetic field $\mathbf{B}$ (Fig. 4-1), is given by

$$e = 2BNlr\omega \sin \omega t = BNA\omega \sin \omega t \qquad (4.2)$$

The second form of (4.2) holds for an arbitrary planar coil of area A. This voltage is available at the slip rings (or brushes), as shown in Fig. 4-1.

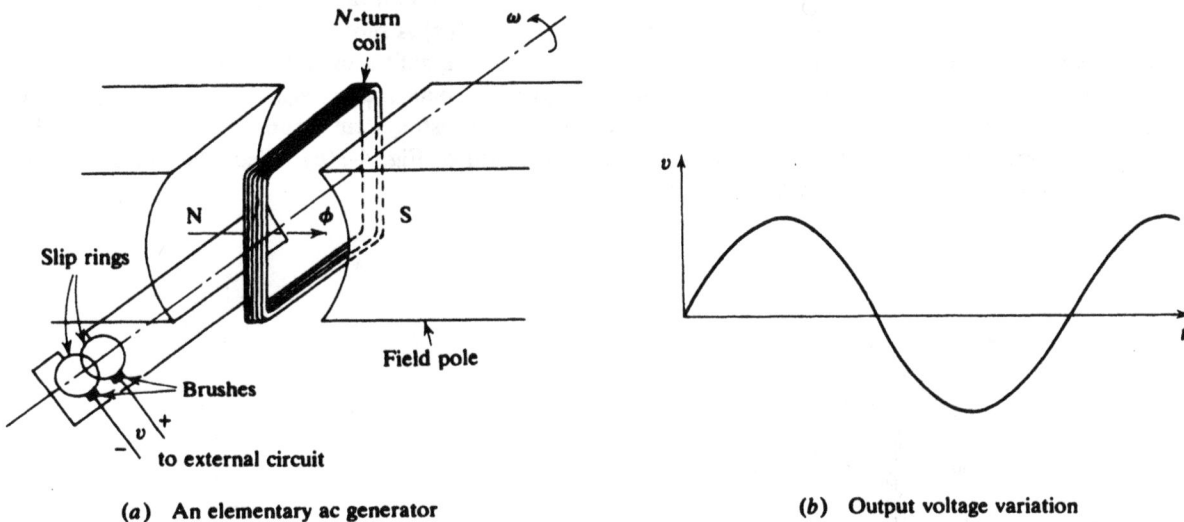

(a) An elementary ac generator

(b) Output voltage variation

Fig. 4-1

The direction of the induced voltage is often determined by the *right-hand rule*, as depicted in Fig. 4-2(a). Clearly, this rule is equivalent to the vector version of (4.1):

$$\text{emf} = \int (l\mathbf{u} \times \mathbf{B}) \cdot d\mathbf{l}$$

Motor action is based on Ampere's law, (1.2), which we rewrite as the *Bli-rule*:

$$F = B(li)_\perp \qquad (4.3)$$

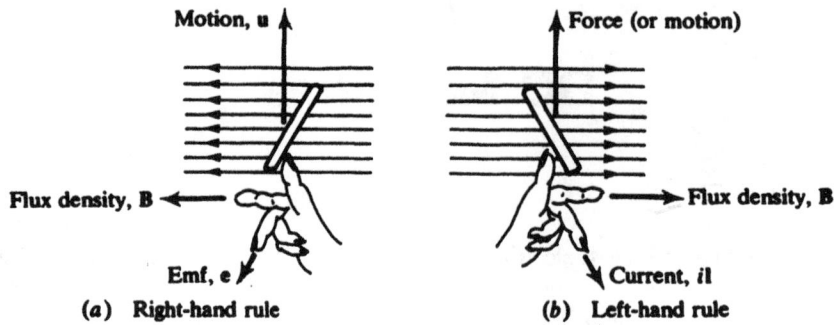

(a) Right-hand rule   (b) Left-hand rule

Fig. 4-2

Here, F is the magnitude of the force on a conductor carrying a directed current element il whose component normal to the uniform magnetic field **B** is $(li)_\perp$. The direction of the force may be obtained by the *left-hand rule*, shown in Fig. 4-2(b).

Just as an ac sinusoidal voltage is produced at the terminals of a generator, the torque produced by the coil fed at the brushes from an ac source would be alternating in nature, with a zero time-average value. If fed from a dc source, the resulting torque will align the coil (in a neutral position) as shown in Fig. 4-1(a). The time-average value of the torque will be zero.

4.2 COMMUTATOR ACTION

In order to get a unidirectional polarity at a brush, or to obtain a unidirectional torque from a coil in a magnetic field, the slip-ring-and-brush mechanism of Fig. 4-1(a) is modified to the one shown in Fig. 4-3(a). Notice that instead of two slip rings we now have one ring split into two halves that are insulated from each other. The brushes slide on these halves, known as *commutator segments*. It can be readily verified by applying the right-hand rule that such a commutator-brush system results in the brushes having definite polarities, corresponding to the output voltage waveform of Fig. 4-3(b). Thus the average output voltage is nonzero and we obtain a dc output at the brushes.

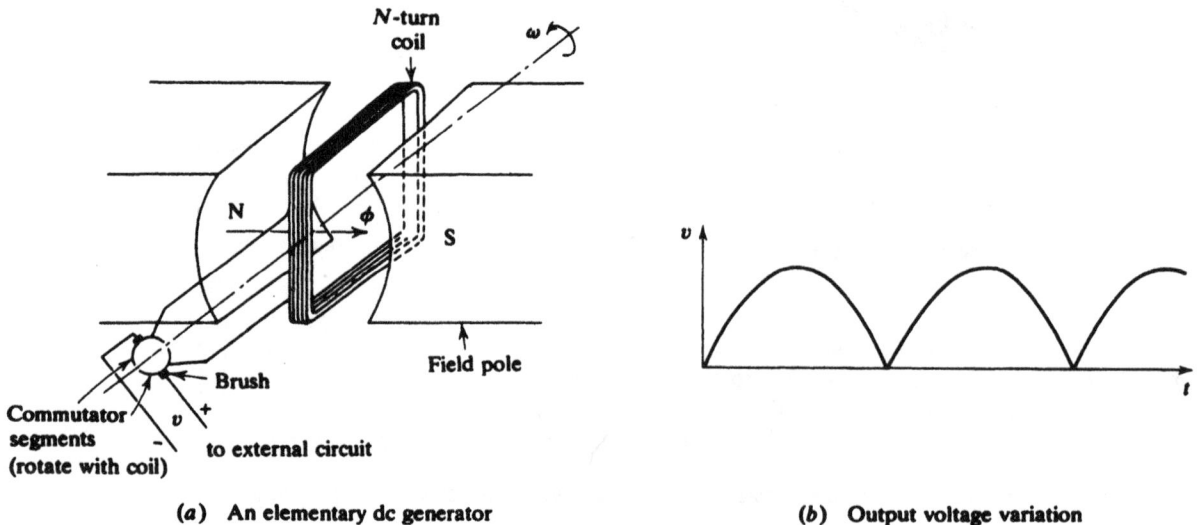

(a) An elementary dc generator   (b) Output voltage variation

Fig. 4-3

It can also be verified, by applying the left-hand rule, that if the coil connected to the commutator-brush system is fed from a dc source, the resulting torque is unidirectional.

The commutator-brush mechanism is an integral part of usual dc machines, the only exception being the *Faraday disk*, or *homopolar machine*. (See Problem 7.24.)

4.3 ARMATURE WINDINGS AND PHYSICAL FEATURES

Figure 4-4 shows some of the important parts and physical features of a dc machine. (For the meaning of GNP and MNP, see Section 4.8.) The *field poles*, which produce the needed flux, are mounted on the *stator* and carry windings called *field windings* or *field coils*. Some machines carry several sets of field windings on the same *pole core*. To facilitate their assembly, the cores of the poles are built of sheet-steel laminations. (Because the field windings carry direct current, it is not electrically necessary to have the cores laminated.) It is, however, necessary for the *pole faces* to be laminated, because of their proximity to the *armature windings*. The *armature core*, which carries the armature windings, is generally on the *rotor* and is made of sheet-steel laminations. The *commutator* is made of hard-drawn copper segments insulated from one another by mica. As shown in Fig. 4-5, the armature windings are connected to the commutator segments over which the carbon *brushes* slide and serve as leads for electrical connection. The armature winding is the load-carrying winding.

The armature winding may be a *lap winding* [Fig. 4-5(a)] or a *wave winding* [Fig. 4-5(b)], and the various coils forming the armature winding may be connected in a series-parallel combination. It is found that in a simplex lap winding the number of paths in parallel, a, is equal to the number of poles, p; whereas in a simplex wave winding the number of parallel paths is always 2.

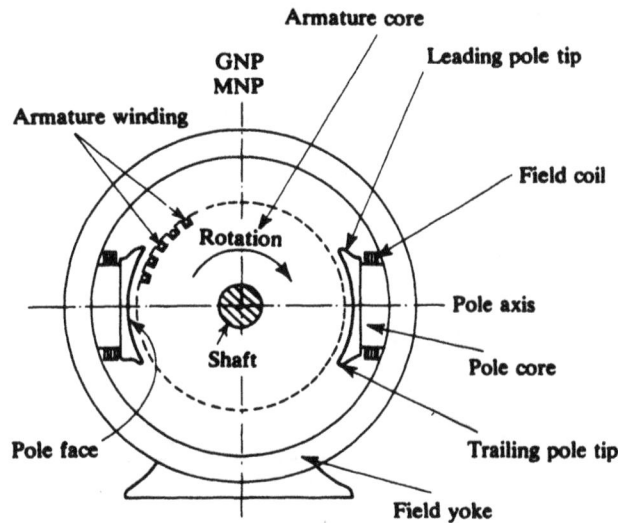

Fig. 4-4

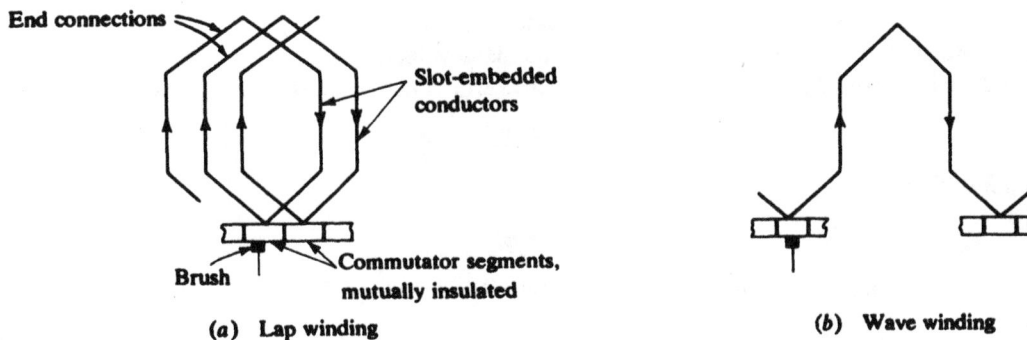

(a) **Lap winding**   (b) **Wave winding**

Fig. 4-5

4.4 EMF EQUATION

Consider a conductor rotating at n rpm in the field of p poles having a flux ϕ per pole. The total flux cut by the conductor in n revolutions is $p\phi n$; hence, the flux cut per second, giving the induced voltage e, is

$$e = \frac{p\phi n}{60} \quad \text{(V)} \qquad (4.4)$$

If there is a total of z conductors on the armature, connected in a parallel paths, then the effective number of conductors in series is z/a, which produce the total voltage E in the armature winding. Hence, for the entire winding, (4.4) gives the *emf equation*:

$$E = \frac{p\phi n}{60} \frac{z}{a} = \frac{zp}{2\pi a} \phi\omega_m \quad \text{(V)} \qquad (4.5)$$

where $\omega_m = 2\pi n/60$ (rad/s). This may also be written as

$$E = k_a\phi\omega_m = k_g\phi n \qquad (4.6)$$

where $k_a \equiv zp/2\pi a$ (a dimensionless constant) and $k_g = zp/60a$. If the magnetic circuit is linear (i.e. if there is no saturation), then

$$\phi = k_f i_f \qquad (4.7)$$

where i_f is the field current and k_f is a proportionality constant; and (4.6) becomes

$$E = ki_f\omega_m \qquad (4.8)$$

where $k \equiv k_f k_a$, a constant. For a nonlinear magnetic circuit, E versus I_f is a nonlinear curve for a given speed, as shown in Fig. 4-6.

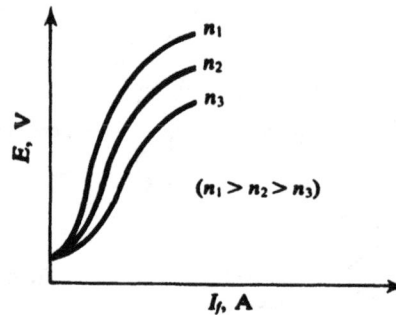

Fig. 4-6

4.5 TORQUE EQUATION

The mechanical power developed by the armature is $T_e\omega_m$, where T_e is the (electromagnetic) torque and ω_m is the armature's angular velocity. If this torque is developed while the armature current is i_a at an armature (induced) voltage E, then the armature power is Ei_a. Thus, ignoring any losses in the armature,

$$T_e\omega_m = Ei_a$$

which becomes, from (4.6),

$$T_e = k_a\phi i_a \qquad (4.9)$$

This is known as the *torque equation*. For a linear magnetic circuit, (4.7) and (4.9) yield

$$T_e = ki_f i_a \qquad (4.10)$$

where $k \equiv k_f k_a$, as in (4.8). Thus, k may be termed the *electromechanical energy-conversion constant*.

Notice that in (4.7) through (4.10) lowercase letters have been used to designate instantaneous values, but that these equations are equally valid under steady state.

4.6 SPEED EQUATION

The armature of a dc motor may be schematically represented as in Fig. 4-7. Under steady state we have

$$V - E = I_a R_a \tag{4.11}$$

Substituting (4.6) in (4.11) yields

$$\omega_m = \frac{V - I_a R_a}{k_a \phi} \tag{4.12}$$

which, for a linear magnetic circuit, becomes

$$\omega_m = \frac{V - I_a R_a}{k I_f} \tag{4.13}$$

An alternate form of (4.13) is

$$n = \frac{V - I_a R_a}{k_m I_f} = \frac{V - I_a R_a}{k_g \phi} \quad \text{(rpm)} \tag{4.14}$$

where $k_m \equiv 2\pi k/60$ ($\Omega \cdot$ min). Equation (4.13) or (4.14) is known as the *speed equation*.

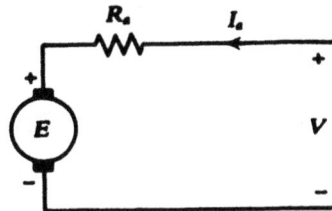

Fig. 4-7

4.7 MACHINE CLASSIFICATION

DC machines may be classified on the basis of the interconnections between the field and armature windings. See Fig. 4-8(a) to (g).

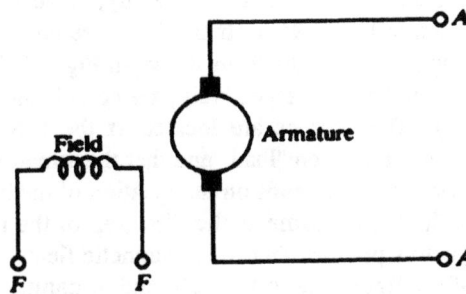

(a) Separately excited

Fig. 4-8(a)

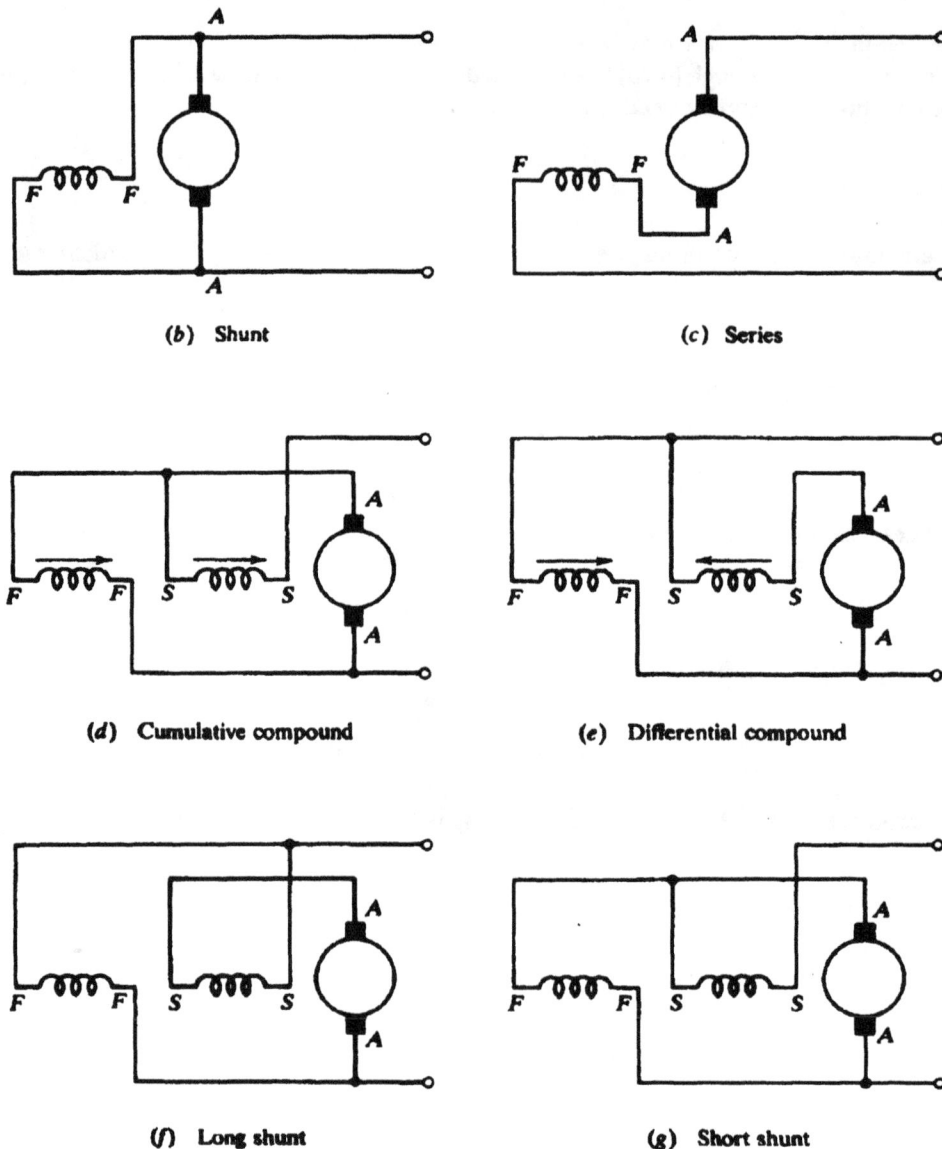

(b)   Shunt   (c)   Series

(d)   Cumulative compound   (e)   Differential compound

(f)   Long shunt   (g)   Short shunt

Fig. 4-8(b) to (g)

4.8 AIRGAP FIELDS AND ARMATURE REACTION

In the discussion so far, we have assumed no interaction between the fields produced by the field windings and by the current-carrying armature windings. In reality, however, the situation is quite different. Consider the two-pole machine shown in Fig. 4-9(a). If the armature does not carry any current (that is, if the machine is on no-load), the airgap field takes the form shown in Fig. 4-9(b). The *geometric neutral plane* and the *magnetic neutral plane* (GNP and MNP, respectively) are coincident. (*Note*: Magnetic lines of force intersect the MNP at right angles.) The brushes are located at the MNP for maximum voltage at the terminals. We now assume that the machine is on "load" and that the armature carries current. The direction of flow of current in the armature conductors depends on the location of the brushes. For the situation in Fig. 4-9(b), the direction of the current flow is the same as the direction of the induced voltages. In any event, the current-carrying armature conductors produce their own magnetic fields, as shown in Fig. 4-9(c), and the airgap field is now the resultant of the fields due to the field and armature windings. This resultant airgap field has the distorted form shown in Fig. 4-9(d). The interaction of the fields due to the armature and field windings is known as *armature reaction*. As a consequence of armature reaction, the airgap field is distorted and the MNP is no longer coincident with the GNP. For maximum voltage at the terminals, the brushes have to be located at the MNP. Thus, one undesirable effect of armature reaction is that the brushes must be shifted constantly, since the deviation of the MNP from the GNP depends on the load (which presumably is always changing).

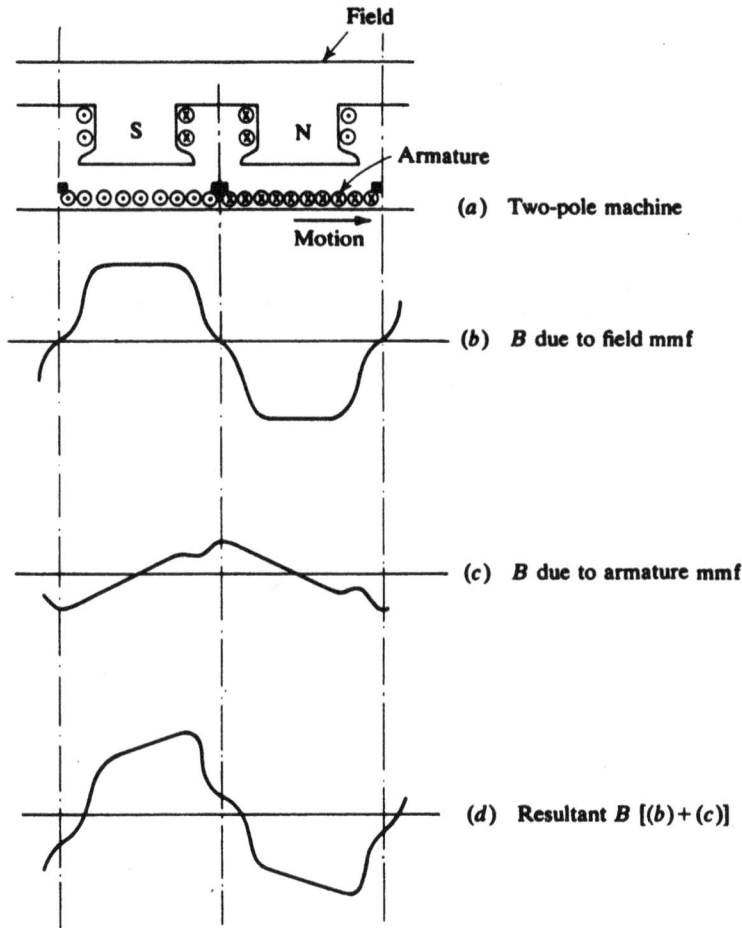

Fig. 4-9

(a) Two-pole machine

(b) B due to field mmf

(c) B due to armature mmf

(d) Resultant B $[(b)+(c)]$

The effect of armature reaction can be analyzed in terms of cross-magnetization and demagnetization, as shown in Fig. 4-10(a). The effect of cross-magnetization can be neutralized by means of *compensating windings*, as shown in Fig. 4-10(b). These are conductors embedded in pole faces, connected in series with the armature windings, and carrying currents in an opposite direction to the currents in the armature conductors that face them [Fig. 4-10(b)]. Once cross-magnetization has been neutralized, the MNP does not shift with load and remains coincident with the GNP at all loads. The effect of demagnetization can be com-

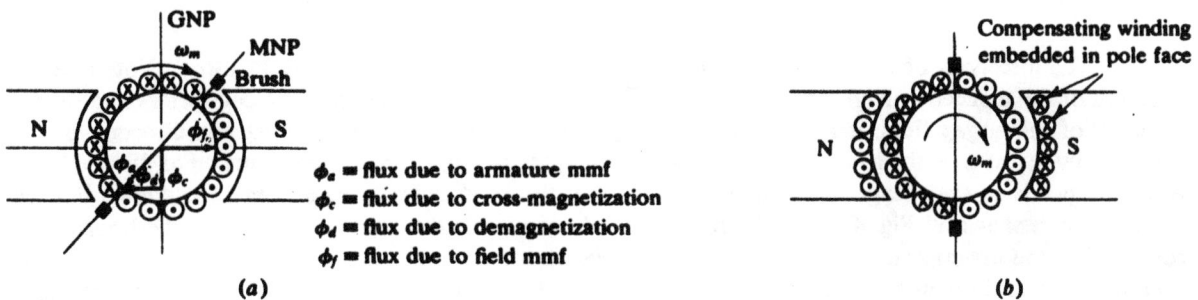

ϕ_a = flux due to armature mmf
ϕ_c = flux due to cross-magnetization
ϕ_d = flux due to demagnetization
ϕ_f = flux due to field mmf

(a)   (b)

Fig. 4-10

pensated for by increasing the mmf on the main field poles. Because the net effect of armature reaction can be neutralized, we are justified in our preceding and succeeding discussions when we assume no "coupling" between the armature and field windings.

4.9 REACTANCE VOLTAGE AND COMMUTATION

In discussing the action of the commutator, we indicated that the direction of flow of current in a coil undergoing commutation reverses by the time the brush moves from one commutator segment to the other. This is schematically represented in Fig. 4-11. The flow of current in coil α for three different instants is shown. We have assumed that the current fed by a commutator segment is proportional to the area of contact between the brush and the segment. Thus, for satisfactory commutation, the direction of flow of current in coil α must completely reverse [Fig. 4-11(a) and (c)] by the time the brush moves from segment 2 to segment 3. The ideal situation is represented by the straight line in Fig. 4-12; it may be termed *straight-line commutation*. Because coil α has some inductance L, the change of current, ΔI, in a time Δt induces a voltage $L(\Delta I/\Delta t)$ in the coil. According to Lenz's law, the direction of this voltage, called *reactance voltage*, is opposite to the change (ΔI) which is causing it. As a result, the current in the coil does not completely reverse by the time the brush moves from one segment to the other. The balance of the "unreversed" current jumps over as a spark from the commutator to the brush, with the result that the commutator wears out from pitting. This departure from ideal commutation is also shown in Fig. 4-12.

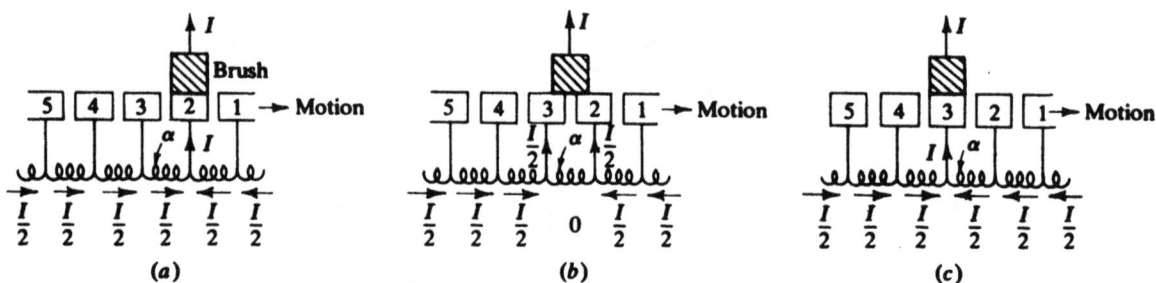

Fig. 4-11

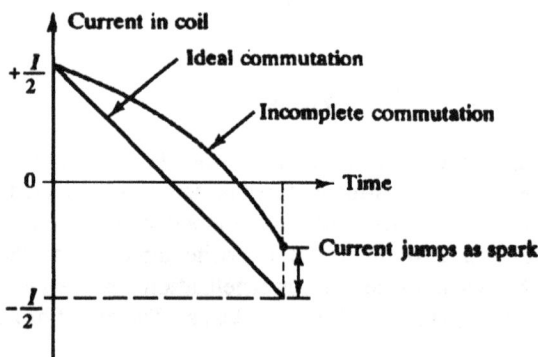

Fig. 4-12

The directions of the current flow and reactance voltage are shown in Fig. 4-13(a). Note that the direction of the induced voltage depends on the direction of rotation of the armature conductors and on the direction of the airgap field; it is given by $\mathbf{u} \times \mathbf{B}$ (or by the right-hand rule). Next, the direction of the current flow depends on the location of the brushes (or tapping points). Finally, the direction of the reactance voltage depends on the change in the direction of current flow and is determined from Lenz's law. For the brush position shown in Fig. 4-13(a), the reactance voltage retards the current reversal. If the brushes are advanced in the direction of rotation (for generator operation), we may notice, from Fig. 4-13(b), that the (rotation-) induced voltage opposes the reactance voltage, so that the current reversal is less impeded than when the reactance voltage acted alone, as in Fig. 4-13(a). We may further observe that the coil undergoing commutation, being near the tip of the south pole, is under the influence of the field of a weak south pole. From this argument, we may conclude that commutation improves if we advance the brushes. But this is not a very practical solution. The same—perhaps better—results can be achieved if we keep the brushes at the GNP, or MNP, as in Fig. 4-13(a), but produce the "field of a weak south pole" by introducing appropriately wound auxiliary poles, called *interpoles* or *commutating poles*. See Fig. 4-13(c).

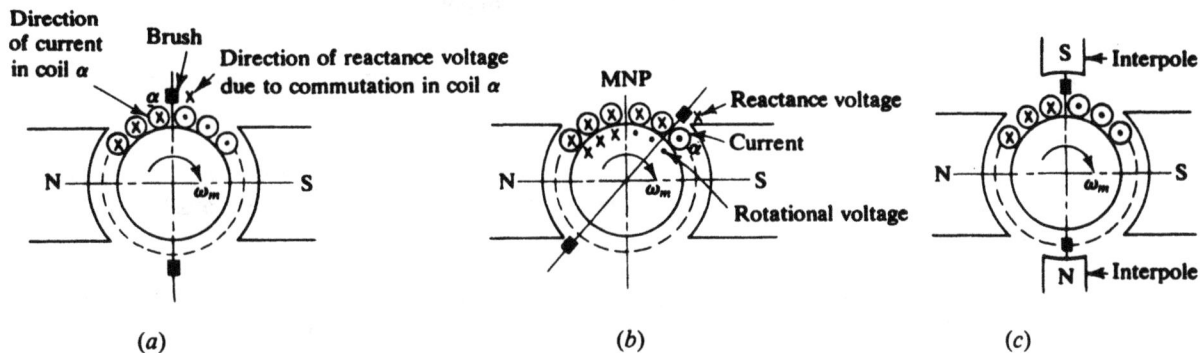

(a)   (b)   (c)

Fig. 4-13

4.10 EFFECT OF SATURATION ON VOLTAGE BUILDUP IN A SHUNT GENERATOR

Saturation plays a very important role in governing the behavior of dc machines. To observe one of its consequences, consider the self-excited shunt generator of Fig. 4-8(b). Under steady state,

$$V = I_f R_f \quad \text{and} \quad E = V + I_a R_a = I_f R_f + I_a R_a$$

These equations are represented by the upper straight lines in Fig. 4-14(a). Notice that the voltages V and E will keep building up and no equilibrium point can be reached. On the other hand, if we include the effect of saturation, as in Fig. 4-14(b), then point P, where the field-resistance line intersects the saturation curves defines the equilibrium.

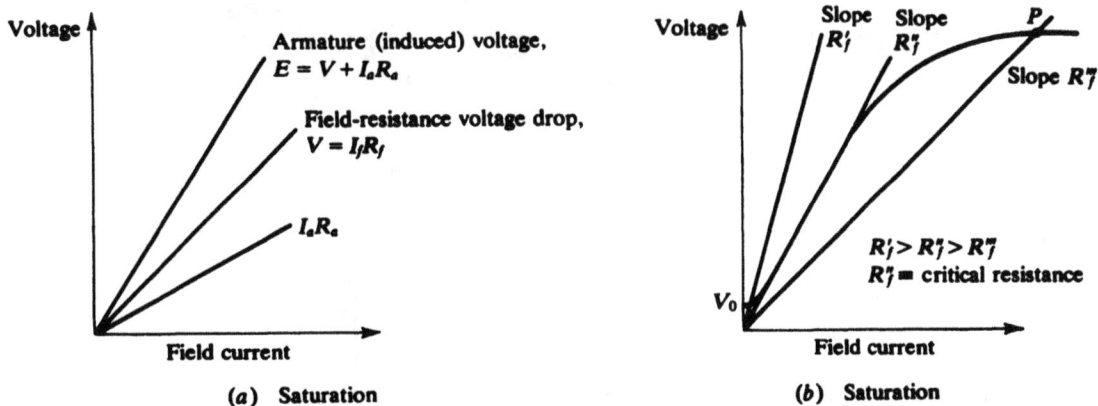

(a) Saturation   (b) Saturation

Fig. 4-14. No-load characteristic of a shunt generator.

Figure 4-14(b) indicates some residual magnetism, as measured by the small voltage V_0. Also indicated in Fig. 4-14(b) is the *critical resistance*: a field resistance greater than the critical resistance (for a given speed) would not let the shunt generator build up an appreciable voltage. Finally, we should ascertain that the polarity of the field winding is such that a current through it produces a flux that aids the residual flux. If instead the two fluxes tend to neutralize, the machine voltage will not build up. To summarize, the conditions for the building up of a voltage in a shunt generator are the presence of residual flux (to provide starting voltage), field-circuit resistance less than the critical resistance, and appropriate polarity of the field winding.

4.11 LOSSES AND EFFICIENCY

Besides the volt-amperage and speed-torque characteristics, the performance of a dc machine is measured by its efficiency:

$$\text{efficiency} \equiv \frac{\text{power output}}{\text{power input}} = \frac{\text{power output}}{\text{power output} + \text{losses}}$$

Efficiency may, therefore, be determined either from load tests or by determination of losses. The various losses are classified as follows.

1. *Electrical.* (*a*) Copper losses in various windings, such as the armature winding and different field windings. (*b*) Loss due to the contact resistance of the brush (with the commutator).

2. *Magnetic.* These are the iron losses and include the hysteresis and eddy-current losses in the various magnetic circuits, primarily the armature core and pole faces.

3. *Mechanical.* These include the bearing-friction, windage, and brush-friction losses.

4. *Stray-load.* These are other load losses not covered above. They are taken as 1 percent of the output (as a rule of thumb).

The power flow in a dc generator or motor is represented in Fig. 4-15, in which T_s denotes the shaft torque.

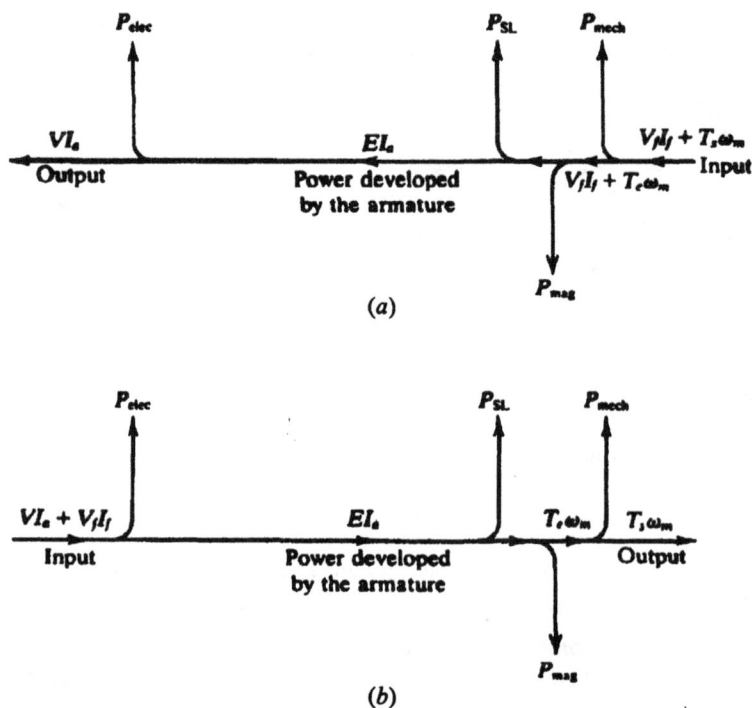

(*a*)

(*b*)

Fig. 4-15

4.12 MOTOR AND GENERATOR CHARACTERISTICS

Load characteristics of motors and generators are usually of greatest interest in determining potential applications of these machines. In some cases (as in Fig. 4-14), no-load characteristics are also of importance.

Typical load characteristics of dc generators are shown in Fig. 4-16, and Fig. 4-17 shows torque speed characteristics of dc motors.

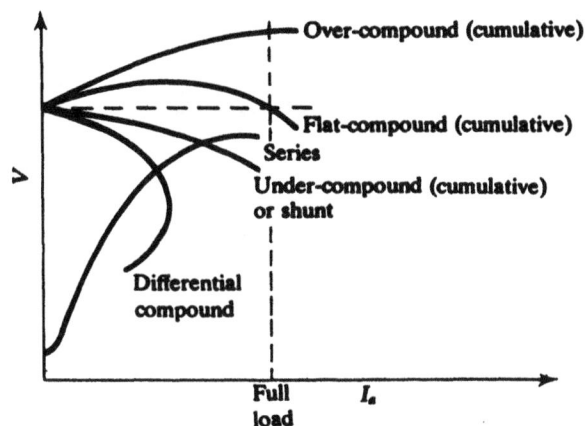

Fig. 4-16. Load characteristics of dc generators.

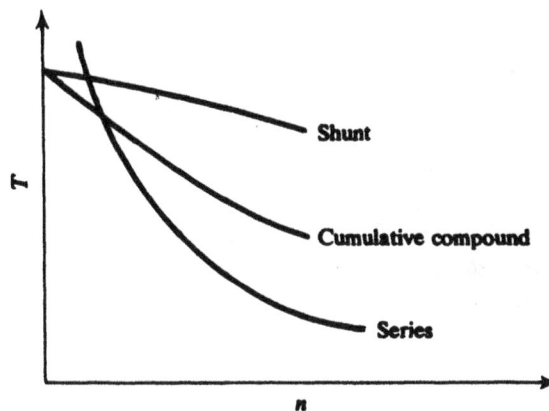

Fig. 4-17. Torque-speed characteristics of dc motors.

4.13 DC MOTOR DYNAMICS

A separately excited motor is represented in Fig. 4-18. For the armature circuit (of an idealized machine) we have

$$v = e + i_a R_a + L_a \frac{di_a}{dt} \tag{4.15}$$

$$e = k i_f \omega_m \tag{4.16}$$

and for the field circuit,

$$v_f = i_f R_f + L_f(i_f) \frac{di_f}{dt} \tag{4.17}$$

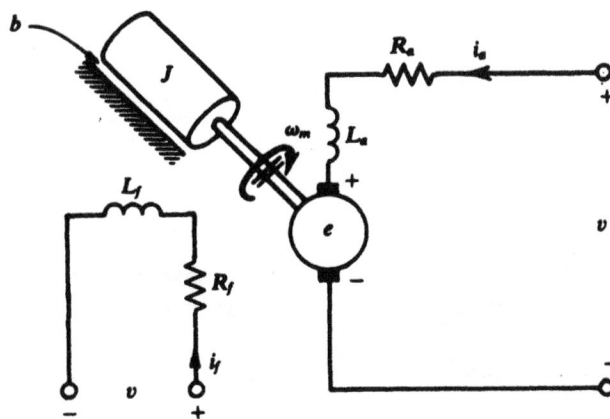

Fig. 4-18

The field-circuit inductance, $L_f(i_f)$, is shown as a nonlinear function of i_f to give generality to the set of equations. This nonlinear function is related to the magnetization curve of the machine or the flux-versus-ampere-turn characteristic of the magnetic circuit of the machine. Summation of torques acting on the motor shaft yields

$$T_e = b\omega_m + J\frac{d\omega_m}{dt} \qquad (4.18)$$

$$T_e = ki_f i_a \qquad (4.19)$$

where b (N · m · s/rad) is a viscous damping coefficient representing mechanical loss and J (kg · m^2) is the moment of inertia of the entire rotating system, including machine rotor, load, couplings, and shaft.

The set of equations (4.15) through (4.19) is nonlinear not only because of the nonlinear coefficients, such as L_f and, possibly, b, but also because of the product terms in (4.16) and (4.19). The set of state equations equivalent to the above set is useful in the analysis of a great number of machine problems. In order to apply these equations, the physical conditions of the specific problem must be introduced in an analytical manner. These conditions include numerical values for the R's, the L's, k, b, and J; descriptions of the input terms, v and v_f; and initial conditions for the state variables. Also, the equations themselves must be modified for different circuit configurations (e.g., for series-field excitation).

Solved Problems

4.1.   Calculate the voltage induced in the armature winding of a 4-pole, lap-wound, dc machine having 728 active conductors and running at 1800 rpm. The flux per pole is 30 mWb.

Because the armature is lap wound, $p = a$, and

$$E = \frac{\phi nz}{60}\left(\frac{p}{a}\right) = \frac{(30 \times 10^{-3})(1800)(728)}{60} = 655.2 \text{ V}$$

4.2.   What is the voltage induced in the armature of the machine of Problem 4.1, if the armature is wave wound?

For a wave-wound armature, $a = 2$. Thus,

$$E = \frac{(30 \times 10^{-3})(1800)(728)}{60}\left(\frac{4}{2}\right) = 1310.4 \text{ V}$$

4.3.   If the armature in Problem 4.1 is designed to carry a maximum line current of 100 A, what is the maximum electromagnetic power developed by the armature?

Because there are 4 parallel paths ($a = p = 4$) in the lap-wound armature, each path can carry a maximum current of

$$\frac{I_a}{a} = \frac{100}{4} = 25 \text{ A}$$

Nevertheless, the power developed by the armature is

$$P_d = EI_a = (655.2)(100) = 65.5 \text{ kW}$$

4.4. By reconnecting the armature of Problem 4.1 in wave, will the developed power be changed?

No. In this case ($a = 2$), the line current is $I_a = 2 \times 25 = 50$ A (25 A being the limit that each path can carry). Hence,

$$P_d = (1310.4)(50) = 65.5 \text{ kW}$$

4.5. Calculate the electromagnetic torque developed by the armature described in Problem 4.1.

From the energy-conversion equation, $EI_a = T_e\omega_m$, and the result of Problem 4.3,

$$T_e = \frac{EI_a}{\omega_m} = \frac{65.5 \times 10^3}{2\pi(1800)/60} = 347.6 \text{ N} \cdot \text{m}$$

4.6. A dc machine has a 4-pole, wave-wound armature with 46 slots and 16 conductors per slot. If the induced voltage in the armature is 480 V at 1200 rpm, determine the flux per pole.

Here $z = 16 \times 46 = 736$, and so, from the emf equation

$$\phi = \frac{60E}{nz}\left(\frac{a}{p}\right) = \frac{(60)(480)}{(1200)(736)}\left(\frac{2}{4}\right) = 16.3 \text{ mWb}$$

4.7. Suppose that in Problem 4.6 the flux per pole remains 16.3 mWb, but the induced voltage measures only 410 V because of flux leakage. Evaluate the *leakage coefficient*, σ, where

$$\sigma \equiv \frac{\text{total flux}}{\text{useful flux}}$$

From the emf equation, the observed voltage is directly proportional to the effective flux. Hence,

$$\sigma = \frac{\text{voltage without leakage}}{\text{voltage with leakage}} = \frac{480}{410} = 1.17$$

4.8. A 4-pole, lap-wound armature has 144 slots with two coil sides per slot, each coil having two turns. If the flux per pole is 20 mWb and the armature rotates at 720 rpm, what is the induced voltage?

Substitute $p = a = 4$, $n = 720$, $\phi = 0.020$, and $z = 144 \times 2 \times 2 = 576$ in the emf equation to obtain

$$E = \frac{(0.020)(720)(576)}{60}\left(\frac{4}{4}\right) = 138.24 \text{ V}$$

4.9. A 10-turn square coil of side 200 mm is mounted on a cylinder 200 mm in diameter. The cylinder rotates at 1800 rpm in a uniform 1.1-T field. Determine the maximum value of the voltage induced in the coil.

From (4.2),

$$E_{max} = BNA\omega = (1.1)(10)(0.200)^2(2\pi \times 1800)/60 = 82.94 \text{ V}$$

4.10. A 100-kW, 230-V, shunt generator has $R_a = 0.05 \ \Omega$ and $R_f = 57.5 \ \Omega$. If the generator operates at rated voltage, calculate the induced voltage at (a) full-load and (b) half full-load. Neglect brush-contact drop.

See Fig. 4-19; $I_f = 230/57.5 = 4$ A.

(a)
$$I_L = \frac{100 \times 10^3}{230} = 434.8 \text{ A}$$

$$I_a = I_L + I_f = 434.8 + 4 = 438.8 \text{ A}$$

$$I_a R_a = (438.8)(0.05) = 22 \text{ V}$$

$$E = V + I_a R_a = 230 + 22 = 252 \text{ V}$$

(b)
$$I_L = 217.4 \text{ A}$$

$$I_a = 217.4 + 4 = 221.4 \text{ A}$$

$$I_a R_a = 11 \text{ V}$$

$$E = 230 + 11 = 241 \text{ V}$$

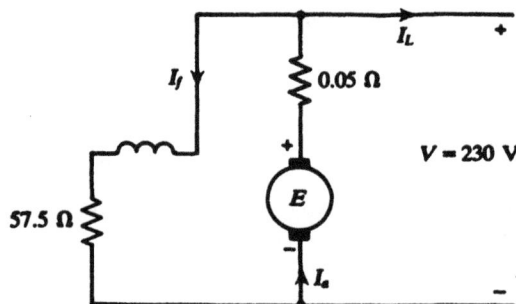

Fig. 4-19

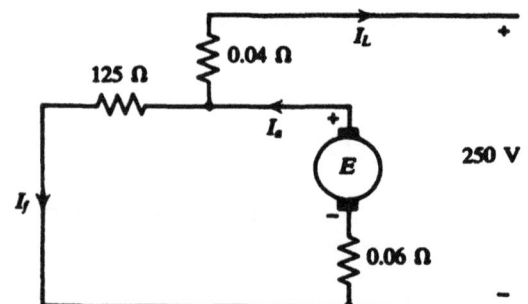

Fig. 4-20

4.11. A 50-kW, 250-V short-shunt compound generator has the following data: $R_a = 0.06 \ \Omega$, $R_{se} = 0.04 \ \Omega$, and $R_f = 125 \ \Omega$. Calculate the induced armature voltage at rated load and terminal voltage. Take 2 V as the total brush-contact drop.

See Fig. 4-20.

$$I_L = \frac{50 \times 10^3}{250} = 200 \text{ A}$$

$$I_L R_{se} = (200)(0.04) = 8 \text{ V}$$

$$V_f = 250 + 8 = 258 \text{ V}$$

$$I_a = 200 + 2.06 = 202.06 \text{ A}$$

$$I_a R_a = (202.06)(0.06) = 12.12 \text{ V}$$

$$E = 250 + 12.12 + 8 + 2 = 272.12 \text{ V}$$

4.12. Repeat Problem 4.11 for a long-shunt compound connection (Fig. 4-21).

$$I_L = 200 \text{ A}$$

$$I_f = \frac{250}{125} = 2 \text{ A}$$

$$I_a = 200 + 2 = 202 \text{ A}$$

$$I_a(R_a + R_{se}) = 202(0.06 + 0.04) = 20.2 \text{ V}$$

$$E = 250 + 20.2 + 2 = 272.2 \text{ V}$$

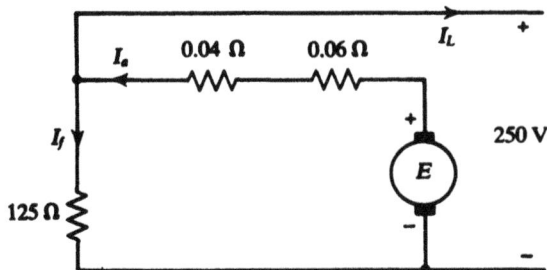

Fig. 4-21

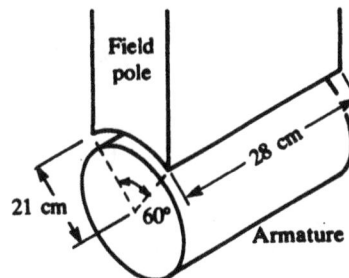

Fig. 4-22

4.13. The generator of Problem 4.10 has 4 poles, is lap wound with 326 armature conductors, and runs at 650 rpm on full-load. If the bore of the machine is 42 cm (in diameter), its axial length is 28 cm, and each pole subtends an angle of 60°, determine the airgap flux density.

A portion of the machine is illustrated in Fig. 4-22. From Problem 4.10,

$$E = 252 = \frac{\phi n z}{60}\left(\frac{p}{a}\right) \quad \text{whence} \quad \phi = 71.35 \text{ mWb}$$

The pole surface area is

$$A = r\theta l = (0.21)(\pi/3)(0.28) = 0.0616 \text{ m}^2$$

Hence
$$B = \frac{\phi}{A} = \frac{71.35 \times 10^{-3}}{0.0616} = 1.16 \text{ T}$$

4.14. A separately excited dc generator has a constant loss of $P_c(W)$, and operates at a voltage V and armature current I_a. The armature resistance is R_a. At what value of I_a is the generator efficiency a maximum?

$$\text{output} = VI_a$$

$$\text{input} = VI_a + I_a^2R_a + P_c$$

$$\text{efficiency } \eta = \frac{VI_a}{VI_a + I_a^2R_a + P_c}$$

For η to be a maximum, $d\eta/dI_a = 0$, or

$$V(VI_a + I_a^2R_a + P_c) - VI_a(V + 2I_aR_a) = 0 \quad \text{or} \quad I_a = \sqrt{\frac{P_c}{R_a}}$$

In other words, the efficiency is maximized when the armature loss, $I_a^2R_a$, equals the constant loss, P_c.

4.15. The generator of Problem 4.10 has a total mechanical and core loss of 1.8 kW. Calculate (a) the generator efficiency at full-load and (b) the horsepower output from the prime mover to drive the generator at this load.

From Problem 4.10, $I_f = 4$ A and $I_a = 438.8$ A, so that

$$I_f^2R_f = (16)(57.5) = 0.92 \text{ kW}$$

$$I_a^2R_a = (438.8)^2(0.05) = 9.63 \text{ kW}$$

and

$$\text{total losses} = 0.92 + 9.63 + 1.8 = 12.35 \text{ kW}$$

(a) $$\text{output} = 100 \text{ kW}$$

$$\text{input} = 100 + 12.35 = 112.35 \text{ kW}$$

$$\text{efficiency} = \frac{100}{112.35} = 89\%$$

(b) $$\text{prime mover output} = \frac{112.35 \times 10^3 \text{ W}}{746 \text{ W/hp}} = 150.6 \text{ hp}$$

4.16. (a) At what load does the generator of Problems 4.10 and 4.15 achieve maximum efficiency? (b) What is the value of this maximum efficiency:

(a)   From Problem 4.15, the constant losses are

$$P_c = 920 + 1800 = 2720 \text{ W}$$

Hence, by Problem 4.14,

$$I_a = \sqrt{\frac{2720}{0.05}} = 233.24 \text{ A}$$

and $\qquad I_L = I_a - I_f = 233.24 - 4 = 229.24 \text{ A}$

(b) $\qquad$ output power $= (229.24)(230) = 52.72 \text{ kW}$

$$I_a^2 R_a = P_c = 2.72 \text{ kW} \quad \text{(by Problem 4.14)}$$

input power $= 52.72 + 2(2.72) = 58.16 \text{ kW}$

maximum efficiency $= \dfrac{52.72}{58.16} \approx 90.6\%$

4.17. The no-load (or saturation) characteristic of a shunt generator at 1200 rpm is shown in Fig. 4-23. The field has 500 turns per pole. (a) Determine the critical field resistance for self-excitation at 1200 rpm. (b) What is the total field-circuit resistance if the induced voltage is 230 V?

From Fig. 4-23:

(a) $\qquad$ critical field resistance $= \dfrac{200}{1500/500} = 66.67 \ \Omega$

(b) $\qquad$ field resistance at 230 V $= \dfrac{230}{2500/500} = 46 \ \Omega$

4.18. Let Fig. 4-23 represent the saturation curve at 1200 rpm of a series motor. The motor has $k_a = 40$ in (4.6), and is wound with 8 turns per pole. The total series-field and armature-circuit resistances are 25 mΩ and 50 mΩ, respectively. Determine the flux per pole at (a) $E = 200$ V, (b) an excitation of 2600 At per pole.

(a) $\qquad$ From (4.6),

$$\phi = \frac{E}{k_a \omega_m} = \frac{200}{40(2\pi \times 1200)/60} = 39.8 \text{ mWb}$$

(b) $\qquad$ At 2600 At per pole, $E \approx 232$ V (from Fig. 4-23). Thus,

$$\phi = \frac{232}{200}(39.8) = 46.15 \text{ mWb}$$

4.19. For a certain load, the motor of Problem 4.18 runs at 200 V while taking 325 A in current. If core loss is 220 W, and friction and windage loss is 40 W, determine (a) electromagnetic torque developed, (b) motor speed, (c) mechanical power output, and (d) motor efficiency.

At $325 \times 8 = 2600$ At per pole, the flux per pole is, from Problem 4.18(b),

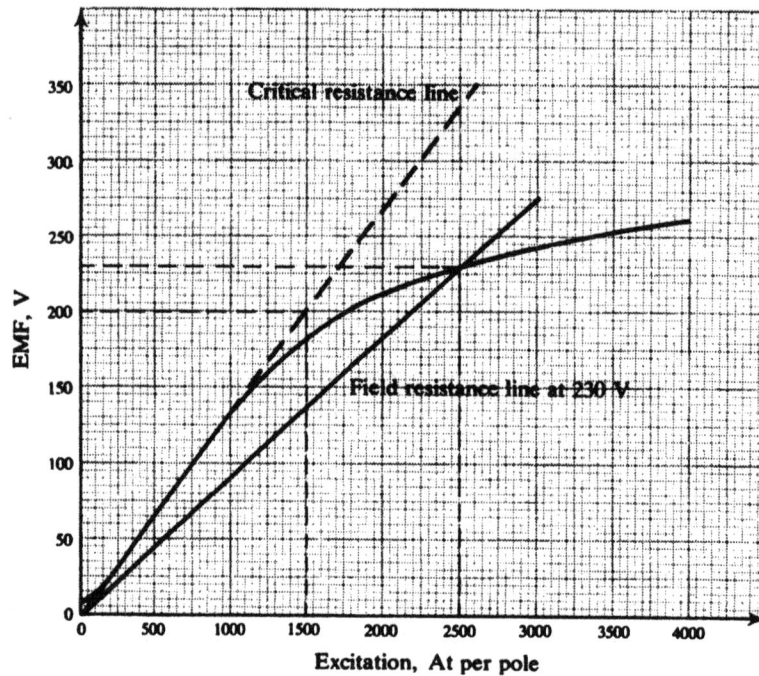

Fig. 4-23

$$\phi = 0.04615 \text{ Wb}$$

(a)
$$T_e = k_a\phi I_a = (40)(0.04615)(325) = 600 \text{ N} \cdot \text{m}$$

(b)
$$E = 200 - (325)(0.025 + 0.050) = 175.6 \text{ V}$$
$$\omega_m = \frac{E}{k_a\phi} = \frac{175.6}{(40)(0.04615)} = 95.14 \text{ rad/s}$$

or $n = 908$ rpm.

(c)
$$\text{output power} = (600)(95.14) = 57.084 \text{ kW}$$

or 76.52 hp.

(d)
$$\text{ohmic loss} = (325)^2(0.025 + 0.050) = 7922 \text{ W}$$
$$\text{core loss} \qquad\qquad\qquad = 220 \text{ W}$$
$$\text{windage and friction loss} \qquad = 40 \text{ W}$$
$$\text{total losses} \qquad\qquad\quad = 8182 \text{ W}$$
$$\text{efficiency} = \frac{57084}{57084 + 8182} = 87.5\%$$

4.20. If, in Problem 4.18, there were no motor saturation, the critical resistance line in Fig. 4-23 might be taken as the motor no-load characteristic at the same speed. Qualitatively, how would the answers to Problem 4.19 be affected?

Flux higher; speed lower, torque higher; power almost unchanged; efficiency higher.

4.21. If the effects of armature reaction effectively canceled 500 At per pole of the main field mmf of the motor of Problem 4.19, describe qualitatively how the answers to Problem 4.19 would be affected. (Assume the original saturated no-load curve.)

Flux lower; speed higher; torque lower; power approximately unchanged; efficiency slightly increased.

4.22. A 20-hp, 250-V shunt motor has an armature-circuit resistance (including brushes and interpoles) of 0.22 Ω and a field resistance of 170 Ω. At no-load and rated voltage, the speed is 1200 rpm and the armature current is 3.0 A. At full-load and rated voltage, the line current is 55 A, and the flux is reduced 6% (due to the effects of armature reaction) from its value at no-load. What is the full-load speed?

$$E_{\text{no-load}} = 250 - (3.0)(0.22) = 249.3 \text{ V}$$

$$I_f = \frac{250}{170} = 1.47 \text{ A}$$

$$E_{\text{full-load}} = 250 - (55 - 1.47)(0.22) = 238.2 \text{ V}$$

From (4.6), ω_m is proportional to E/ϕ; whence

$$n_{m,\text{full-load}} = 1200 \left(\frac{238.2}{249.3}\right)\left(\frac{1}{0.94}\right) = 1220 \text{ rpm}$$

4.23. A 230-V shunt motor has the no-load (zero armature current) magnetization characteristic at 1800 rpm shown in Fig. 4-23. The full-load armature current is 100 A; the armature-circuit resistance (including brushes and interpoles) is 0.12 Ω. (a) The motor runs at 1800 rpm at full-load and also at no-load. Determine the demagnetizing effect of armature reaction at full-load, in At per pole. (b) A long-shunt, cumulative, series-field winding having 8 turns per pole and a resistance of 0.08 Ω is added to the motor. Determine the speed at full-load current and rated voltage.

(a) $$E_{\text{full-load}} = 230 - (100)(0.12) = 218 \text{ V}$$

Recall the speed equation, (4.12). Since $V - I_a R_a$ decreases in going from no-load to full-load, the mmf per pole must decrease by a proportional amount for ω_m to remain unchanged:

$$\text{mmf}_{\text{full-load}} = \frac{218}{230}(2500) = 2370 \text{ At per pole}$$

(assuming magnetic linearity in this range) and

$$\text{decrease in mmf} = 2500 - 2370 = 130 \text{ At per pole}$$

(b) $$E_{\text{full-load}} = 230 - (100)(0.12 + 0.08) = 210 \text{ V}$$

$$\text{mmf}_{\text{full-load}} = 2370 + (100)(8) = 3170 \text{ At per pole}$$

This mmf generates an emf of ≈ 247 V (from Fig. 4-23). Since n is proportional to E/ϕ or E/mmf,

$$n_2 = 1800 \left(\frac{210}{218}\right)\left(\frac{2370}{3170}\right) = 1296 \text{ rpm}$$

4.24. A 10-hp, 230-V shunt motor takes a full-load line current of 40 A. The armature and field resistances are 0.25 Ω and 230 Ω, respectively. The total brush-contact drop is 2 V and the core and friction losses are 380 W. Calculate the efficiency of the motor. Assume that stray-load loss is 1% of output.

$$\text{input} = (40)(230) = 9200 \text{ W}$$

$$\text{field-resistance loss} = \left(\frac{230}{230}\right)^2 (230) = 230 \text{ W}$$

$$\text{armature-resistance loss} = (40-1)^2(0.25) = 380 \text{ W}$$

$$\text{core loss and friction loss} = 380 \text{ W}$$

$$\text{brush-contact loss} = (2)(39) = 78 \text{ W}$$

$$\text{stray-load loss} = \frac{10}{100} \times 746 = 78 \text{ W}$$

$$\text{total losses} = 1143 \text{ W}$$

$$\text{power output} = 9200 - 1143 = 8057 \text{ W}$$

$$\text{efficiency} = \frac{8057}{9200} = 87.6\%$$

4.25. A 230-V shunt motor delivers 30 hp at the shaft at 1120 rpm. If the motor has an efficiency of 87% at this load, determine (a) the total input power and (b) the line current. (c) If the torque lost due to friction and windage is 7% of the shaft torque, calculate the developed torque.

(a) $$\text{input power} = \frac{\text{output}}{\text{efficiency}} = \frac{(30)(746)}{0.87} = 25.72 \text{ kW}$$

(b) $$\text{input current} = \frac{\text{input power}}{\text{input voltage}} = \frac{25720}{230} = 111.8 \text{ A}$$

(c) $$\text{output torque} = \frac{\text{output power}}{\text{angular velocity}} = \frac{(30)(746)}{(2\pi \times 1120)/60} = 190.8 \text{ N} \cdot \text{m}$$

$$\text{developed torque} = (1.07)(190.8) = 204.2 \text{ N} \cdot \text{m}$$

4.26. A 10-kW, 250-V shunt generator, having an armature resistance of 0.1 Ω and a field resistance of 250 Ω, delivers full-load at rated voltage and 800 rpm. The machine is now run as a motor while taking 10 kW at 250 V. What is the speed of the motor? Neglect brush-contact drop.

As a generator:

$$I_f = \frac{250}{250} = 1 \text{ A} \qquad I_L = \frac{10 \times 10^3}{250} = 40 \text{ A}$$

$$I_a = 40 + 1 = 41 \text{ A} \qquad I_a R_a = (41)(0.1) = 4.1 \text{ V}$$

$$E_g = 250 + 4.1 = 254.1 \text{ V}$$

As a motor:

$$I_L = \frac{10 \times 10^3}{250} = 40 \text{ A} \qquad I_f = \frac{250}{250} = 1 \text{ A}$$

$$I_a = 40 - 1 = 39 \text{ A} \qquad I_a R_a = (39)(0.1) = 3.9 \text{ V}$$

$$E_m = 250 - 3.9 = 246.1 \text{ V}$$

Now

$$\frac{n_m}{n_g} = \frac{E_m}{E_g} \quad \text{or} \quad n_m = \frac{E_m}{E_g} n_g = \frac{246.1}{254.1} (800) = 774.8 \text{ rpm}$$

4.27. Figure 4-24 depicts the *Ward-Leonard system* for controlling the speed of the motor M. The generator field voltage, v_{fg}, is the input and the motor speed, ω_m, is the output. Obtain an expression for the transfer function for the system, assuming idealized machines. The load on the motor is given by $J\dot{\omega}_m + b\omega_m$, and the generator runs at constant angular velocity ω_g.

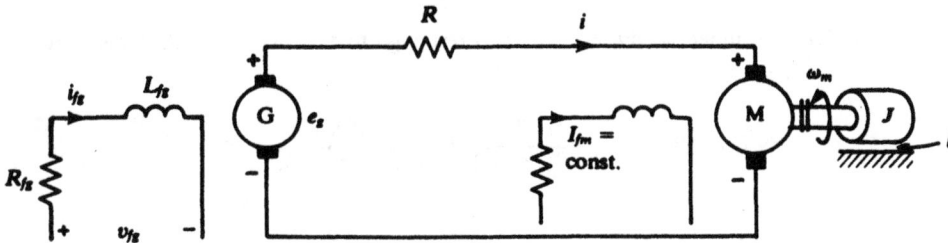

Fig. 4-24

From Fig. 4-24, (*4.16*), and (*4.19*), the equations of motion are:

$$v_{fg} = R_{fg} i_{fg} + L_{fg} \frac{di_{fg}}{dt} \quad \text{or} \quad V_{fg} = (R_{fg} + L_{fg}s) I_{fg}$$

$$e_g = k_g \omega_g i_{fg} = Ri + k_m I_{fm} \omega_m \quad \text{or} \quad k_g \omega_g I_{fg} = RI + k_m I_{fm} \Omega_m$$

$$T_m = k_m I_{fm} i = J\dot{\omega}_m + b\omega_m \quad \text{or} \quad k_m I_{fm} I = (b + Js)\Omega_m$$

Hence,

$$G(s) \equiv \frac{\Omega_m(s)}{V_{fg}(s)} = \frac{k_g \omega_g k_m I_{fm}}{(R_{fg} + L_{fg}s)(k_m^2 I_{fm}^2 + bR + JRs)}$$

4.28. A separately excited generator can be treated as a power amplifier when driven at a constant angular velocity ω_m. Derive an expression for the voltage gain, $V_L(s)/V_f(s)$, in terms of the parameters given in Fig. 4-25 and the proportionality constant $k\omega_m$ in (4.8).

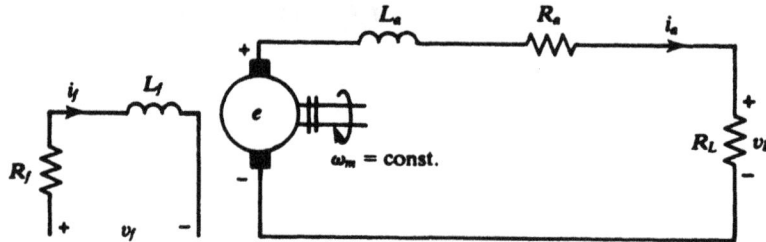

Fig. 4-25

In the transform domain, the equations are, from Fig. 4-25,

$$V_f = (R_f + L_f s)I_f \qquad \mathscr{E} = k\omega_m I_f$$

Also, $\mathscr{E} = (R_a + R_L + L_a s)I_a$ and $V_L + R_L I_a$. Consequently,

$$\frac{V_L(s)}{V_f(s)} = \frac{R_L k\omega_m}{R_f(R_a + R_L)} \cdot \frac{1}{(1 + \tau_f s)(1 + \tau_a s)}$$

where $\tau_f \equiv L_f/R_f$ and $\tau_a \equiv L_a/(R_a + R_L)$.

4.29. In Problem 4.28, $R_a = 0.1\ \Omega$, $R_f = 10\ \Omega$, $R_L = 0.5\ \Omega$, and $k\omega_m = 65$ V/A. What are (a) the voltage gain, and (b) the power gain if the generator is operating under steady state with 25 V applied across the field?

Notice that under steady state the d/dt terms go to zero, i.e., $s \to 0$. Therefore:

(a)
$$\text{voltage gain} = \frac{R_L k\omega_m}{R_f(R_a + R_L)} = \frac{(0.5)(65)}{(10)(0.1 + 0.5)} = 5.42$$

(b)
$$\text{input power to the field} = \frac{(25)^2}{10} = 62.5\ \text{W}$$

$$E = k\omega_m I_f = (65)\left(\frac{25}{20}\right) = 162.5\ \text{V}$$

$$I_a = \frac{162.5}{0.1 + 0.5} = 270.8\ \text{A}$$

$$\text{output power} = (270.8)^2(0.5) = 36\,675\ \text{W}$$

$$\text{power gain} = \frac{36\,675}{62.5} = 587$$

4.30. A separately excited dc motor, having a constant field current, accelerates a pure inertia load from rest. Represent the system by an electrical equivalent circuit. The various symbols are defined in Fig. 4-26(a).

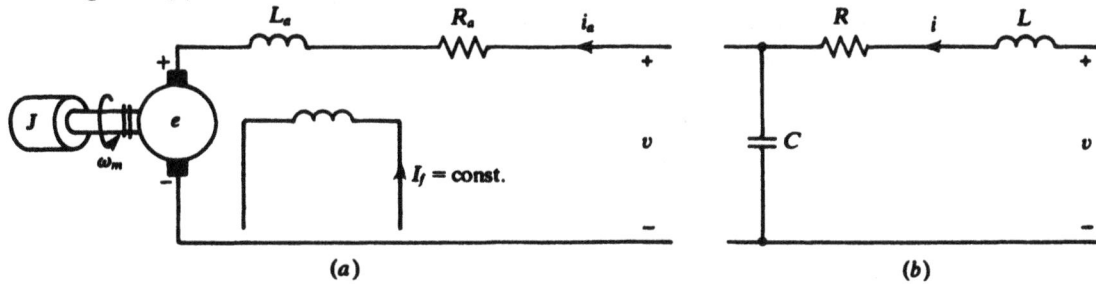

(a)   (b)

Fig. 4-26

The equations of motion are:

$$v = R_a i_a + L_a \frac{di_a}{dt} + e$$

$$e = kI_f \omega_m$$

$$T_e = kI_f i_a = J\dot{\omega}_m$$

These equations yield

$$v = R_a i_a + L_a \frac{di_a}{dt} + \frac{(kI_f)^2}{J} \int i_a \, dt$$

which is similar to

$$v = Ri + L \frac{di}{dt} + \frac{1}{C} \int i \, dt$$

corresponding to the circuit of Fig. 4-26(b). For equivalence: $R \leftrightarrow R_a$, $L \leftrightarrow L_a$, and $C \leftrightarrow J/(kI_f)^2$.

Supplementary Problems

4.31. Derive (*4.2*) from (*2.1*).

4.32. A 6-pole, lap-wound, dc generator armature has 720 active conductors. The generator is designed to generate 420 V at 1720 rpm. Determine the flux per pole. *Ans.* 20.35 mWb

4.33. The armature of the generator of Problem 4.32 is reconnected as wave wound. At what speed must the generator operate to induce 630 V in the armature? *Ans.* 860 rpm

4.34. At what speed, in rpm, must the armature of a dc machine run to develop 572.4 kW at a torque of 4605 N · m?   *Ans.* 1187 rpm

4.35. The armature of a dc machine running at 1200 rpm carries 45 A in current. If the induced armature voltage is 130 V, what is the torque developed by the armature?   *Ans.* 46.5 N · m

4.36. A series generator has the saturation characteristic shown in Fig. 4-23 and has 8 turns per pole. Determine (*a*) the load resistance at which the generator will operate at 220 V, (*b*) the armature current at this load.   *Ans.* (*a*) ≈0.8 Ω (*b*) ≈272 A

4.37. A self-excited shunt generator has the saturation characteristic of Fig. 4-27. (*a*) Find the value of the critical field resistance (above which the generator will not build up). (*b*) What is the no-load terminal voltage if the field-circuit resistance is 50 Ω?   *Ans.* (*a*) 56 Ω; (*b*) 250 V

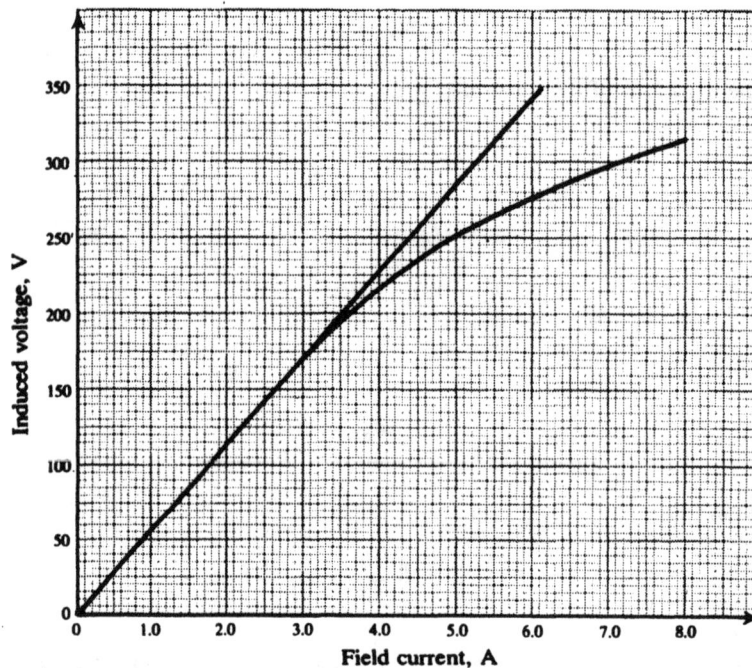

Fig. 4-27

4.38. A self-excited shunt generator supplies a load of 12.5 kW at 125 V. The field resistance is 25 Ω and the armature resistance is 0.1 Ω. The total voltage drop because of brush contact and armature reaction at this load is 3.5 V. Calculate the induced armature voltage.   *Ans.* 139 V

4.39. From Fig. 4-23, determine the minimum speed at which the shunt generator will self-excite if the field-circuit resistance is 66.67 Ω and the field winding has 500 turns per pole.   *Ans.* 1200 rpm

4.40. A 6-pole, lap-wound armature, having 720 conductors, rotates in field of 20.35 mWb. (*a*) If the armature current is 78 A, what is the torque developed by the armature? (*b*) If the induced armature voltage is 420 V, what is the motor speed?   *Ans.* (*a*) 181.9 N · m; (*b*) 1720 rpm

4.41.   A separately excited motor runs at 1045 rpm, with a constant field current, while taking an armature current of 50 A at 120 V. The armature resistance is 0.1 Ω. If the load on the motor changes such that it now takes 95 A at 120 V, determine the motor speed at this load.   *Ans.* 1004 rpm

4.42.   A long-shunt compound generator supplies 50 kW at 230 V. The total field- and armature-circuit resistances are 46 Ω and 0.03 Ω, respectively. The brush-contact drop is 2 V. Determine the percent voltage regulation. Neglect armature reaction.   *Ans.* 3.77%

4.43.   In Problem 4.42, if the series-field resistance is 0.02 Ω, the armature-circuit resistance is 0.01 Ω, and the connection is short-shunt, what is the voltage regulation?   *Ans.* 3.72%

4.44.   A separately excited dc generator has the following data: armature resistance, 0.04 Ω; field resistance, 110 Ω; total core and mechanical losses, 960 W; voltage across the field, 230 V. The generator supplies a load at a terminal voltage of 230 V. Calculate (*a*) the armature current at which the generator has a maximum efficiency, and (*b*) the maximum value of the generator efficiency.
Ans. (*a*) 189.8 A; (*b*) 93.8%

4.45.   A shunt motor operates at a flux of 25 mWb per pole, is lap wound, and has 2 poles and 360 conductors. The armature resistance is 0.12 Ω and the motor is designed to operate at 115 V, taking 60 A in armature current at full-load. (*a*) Determine the value of the external resistance to be inserted in the armature circuit so that the armature current shall not exceed twice its full-load value at starting. (*b*) When the motor has reached a speed of 400 rpm, the external resistance is cut by 50%. What is the armature current then, at this speed? (*c*) The external resistance is completely cut out when the motor reaches its final speed; armature current is then at its full-load value. Calculate the motor speed.
Ans. (*a*) 0.838 Ω; (*b*) 102 A; (*c*) 718.6 rpm

4.46.   Compute the developed torque in parts (*a*), (*b*), and (*c*) of Problem 4.45.
Ans. (*a*) 172 N · m; (*b*) 146 N · m; (*c*) 80 N · m

4.47.   A 230-V shunt motor, having an armature resistance of 0.05 Ω and a field resistance of 75 Ω, draws a line current of 7 A while running light at 1120 rpm. The line current at a given load is 46 A. Determine (*a*) the motor speed at the given load, (*b*) motor efficiency, and (*c*) total core and mechanical losses.
Ans. (*a*) 1110.5 rpm; (*b*) 83.9%; (*c*) 903.9 W

4.48.   If the field-circuit resistance of the motor of Problem 4.47 is increased to 100 Ω at the given load, other conditions remaining unchanged, what is the new speed of the motor?   *Ans.* 1480 rpm

4.49.   The magnetic characteristic of a certain dc motor is the straight line defined by the equation $\phi = 0.001 I_f$, where ϕ (Wb) is the flux per pole and I_f (A) is the field current. The motor is separately excited; $R_a = 0.05 \Omega$; $k_a = 100$. With I_f equal to 10 A and with 400 V applied to the armature terminals, the motor runs at 3000 rpm. Determine I_a, E, and the electromagnetic torque. Neglect armature reaction; R_a is the total armature-circuit resistance.   *Ans.* 1720 A; 314 V; 1720 N · m

4.50.   A dc series motor is connected to a constant-torque load, which may be considered to require a constant electromagnetic torque regardless of motor speed. Neglect the voltage drops due to the armature and series-field resistances, the armature reaction, and the effects of saturation. (*a*) By what percent is the motor speed changed when the line voltage is reduced from 230 V to 200 V? (*b*) Repeat (*a*), assuming a shunt motor. (*c*) State briefly what effects saturation might have on the answers to (*a*) and (*b*).
Ans. (*a*) 13%; (*b*) ≈ 0

4.51. A shunt motor runs at 1100 rpm, at 230 V, and draws a line current of 40 A. The output power (at the shaft) is 10.8 hp. The various losses are core loss, 200 W; friction and windage loss, 180 W; electrical loss due to brush contact, 37 W; stray-load loss, 37 W. The armature- and field-circuit resistances are 0.25 Ω and 230 Ω, respectively. Calculate (a) the motor efficiency, and (b) the speed if the output power is reduced by 50%.   *Ans.* (a) 88.69%; (b) 1125 rpm

4.52. A 4-pole motor is lap wound with 728 conductors, and has a flux of 25 mWb per pole. The armature takes 50 A in current; the resultant demagnetizing effect due to armature reaction reduces the airgap flux by 5%. Calculate the developed torque.   *Ans.* 137.6 N $\cdot$ m

4.53. If the armature of the motor of Problem 4.52 is wave wound, all other data except the armature current remaining unchanged, verify that the developed torque will be unchanged.

4.54. A series motor, having an armature resistance of 0.1 Ω and a field resistance of 0.15 Ω, takes 48 A at 230 V and 720 rpm. The total core and friction losses are 650 W. Neglecting stray-load and brush-contact losses, calculate (a) the developed torque, (b) output power at the shaft, and (c) the motor efficiency.   *Ans.* (a) 138.8 N $\cdot$ m; (b) 13.16 hp; (c) 88.9%

4.55. The load on the motor of Problem 4.54 is reduced so that the motor now takes 32 A. Find (a) the motor speed, (b) the percentage change in the torque. Neglect saturation.   *Ans.* (a) 1100 rpm; (b) 55.6%

4.56. A 250-V, 10 hp shunt motor has an armature resistance of 0.5 Ω and a field resistance of 250 Ω. The motor takes a current of 5 A on no-load and 37.1 A on rated load. Determine its rated-load efficiency.   *Ans.* 79.7%

4.57. Referring to the speed equation of a dc motor, state the possible means by which the speed of the motor may be varied.

4.58. The motor of Fig. 4-26(a) is started from rest with constant field current, I_f. Neglecting L_a, show that by the time the motor reaches its final speed the energy dissipated in the resistance R_a is equal to the energy stored in the rotating parts. What is the numerical value of this energy if $V = 120$ V, $R_a = 0.1$ Ω, kI_f (motor torque constant) = 4 N $\cdot$ m/A, and $J = 40$ kg $\cdot$ m^2?   *Ans.* 18 kJ

4.59. A separately excited dc generator, running at a constant speed, supplies a load having a 0.9 Ω resistance in series with a 1-H inductance. The armature resistance is 0.1 Ω; its inductance is negligible. The field, having a resistance of 50 Ω and an inductance of 5 H, is suddenly connected to a 100-V source. Determine the armature current buildup if the generator voltage constant is $k\omega_m = 40$ V/A.

Ans. $i_a(t) = 80 \left(1 - \dfrac{10}{9} e^{-t} + \dfrac{1}{9} e^{-10t} \right)$ A

4.60. A separately excited motor carries a load given by $300\dot{\omega}_m + \omega_m$ (N $\cdot$ m). The armature resistance is 1 Ω and its inductance is negligible. If 100 V is suddenly applied across the armature, while the field current is constant at I_f, obtain an expression for the motor speed buildup. The motor torque constant is $kI_f = 7$ N $\cdot$ m/A.   *Ans.* $\omega_m(t) = 14(1 - e^{-t/6})$ (rad/s)

4.61. A 220-V dc shunt motor has an armature resistance of 0.2 Ω and a field resistance of 110 Ω. At no-load, the speed is 1000 rpm and the line current is 7 A. At full-load the input power is 11 kW. (a) Calculate

the speed at full-load. (b) Also determine the full-load developed torque.
Ans. (a) 960.7 rpm; (b) 100.38 N · m

4.62. An 8-pole wave wound dc shunt generator, with 778 conductors, while running at 500 rpm supplies a 12.5 Ω resistive load at 250 V. The armature resistance is 0.24 Ω and the field resistance is 250 Ω. Calculate the (a) armature current and (b) flux per pole. *Ans.* (a) 21.0 A; (b) 9.83 mWb

4.63. A 200-V dc shunt motor draws 4 A line current on no-load while running at 700 rpm. The field and armature circuit resistances are 100 Ω and 0.6 Ω, respectively. On load the motor takes 8 kW from the source. For this load determine the (a) speed and (b) developed torque. *Ans.* (a) 624 rpm; (b) 103 N · m

4.64. A lap-connected separately excited dc generator has 728 conductors and 20 mWb flux per pole. The armature resistance is 0.1 Ω. At what speed must the generator run to maintain 220 V at the terminals while supplying 50 A current to a resistive load? *Ans.* 927.2 rpm

4.65. A separately excited dc motor develops 700 N · m torque while drawing 35-A armature current. (a) What is the torque at 70-A input current? (b) Calculate the motor speed in revolutions per minute (rpm) if the back emf at 35-A armature current is 200 V. *Ans.* (a) 1400 N · m; (b) 95.5 rpm

Chapter 5

Polyphase Induction Motors

5.1 GENERAL REMARKS

The induction motor is probably the most common of all motors. Like the dc machine, an induction motor consists of a stator and a rotor, the latter mounted on bearings and separated from the stator by an airgap. The stator core, made up of punchings (or laminations), carries slot-embedded conductors. These conductors are interconnected in a predetermined fashion and constitute the armature windings.

Alternating current is supplied to the stator windings, and the currents in the rotor windings are induced by the magnetic field of the stator currents. The rotor of the induction machine is cylindrical and carries either (1) conducting bars short-circuited at both ends by conducting rings, as in a *cage-type machine* (Fig. 5-1); or (2) a polyphase winding with terminals brought out to slip rings for external connections, as in a *wound-rotor machine* (Fig. 5-2). A wound-rotor winding is similar to that of the stator. Sometimes the cage-type machine is called a *brushless machine* and the wound-rotor machine is termed a *slip-ring machine*.

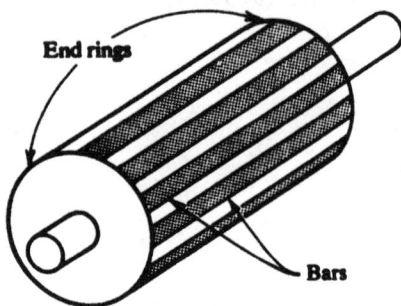

Fig. 5-1

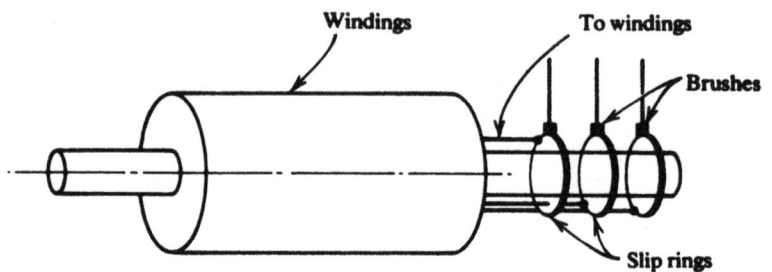

Fig. 5-2

An induction motor operates on the basis of interaction of induced rotor currents and the airgap field. If the rotor is allowed to run under the torque developed by this interaction, the machine will operate as a motor. On the other hand, the rotor may be driven by an external agency beyond a speed such that the machine begins to deliver electric power; it then operates as an induction generator. Almost invariably, induction machines are used as motors.

5.2 MMFs of ARMATURE WINDINGS

As in a dc machine, there are often several independent sets of windings on the stator of an induction motor. For instance, a *three-phase winding* is shown in Fig. 5-3, where each slot contains two coil sides. Such a winding is a *double-layer winding*. Notice also from Fig. 5-3 that it is a four-pole winding, but that the *pole pitch* (of nine teeth) is slightly greater than the *coil pitch* (of eight teeth). Thus, it is a *fractional-pitch* (or *chorded*) *winding*. Finally, observe that in this case there are three slots per pole per phase. If the number of slots per pole per phase were nonintegral, the winding would be known as a *fractional-slot winding*.

Because the armature winding consists of interconnected coils, it is advantageous to consider the mmf of a single full-pitch coil having N turns. From Fig. 5-4(a) it is evident that the machine has two poles; and from Ampere's circuital law it follows that the mmf has the uniform value Ni (At) between the coil sides,

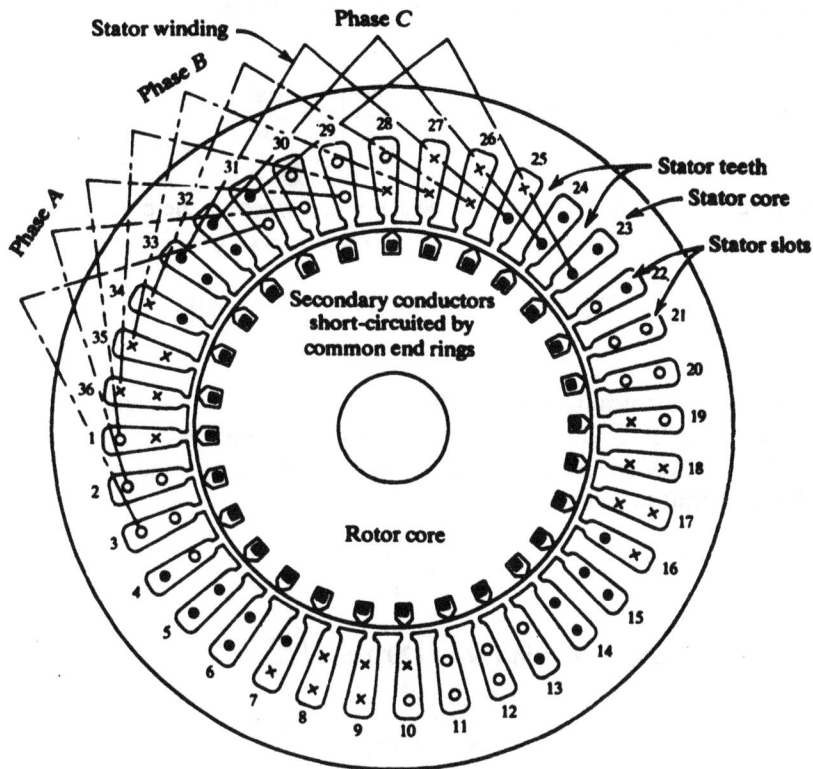

Key: O, phase A; ×, phase B; ●, phase C.

Fig. 5-3

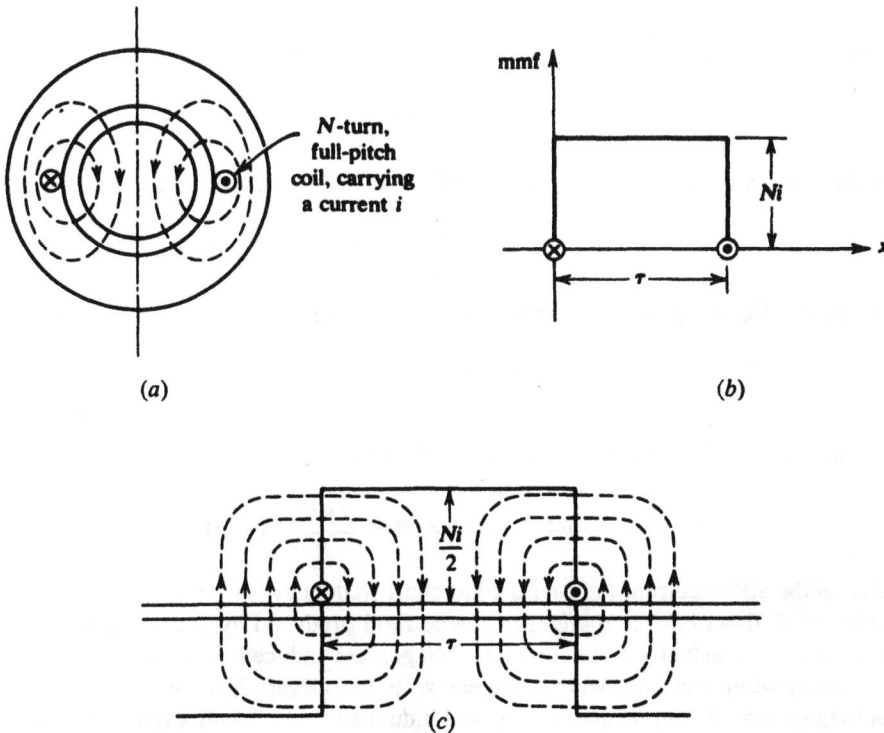

(a)

(b)

(c)

Fig. 5-4

as depicted in Fig. 5-4(b). Thus, the mmf per pole is $Ni/2$, as indicated in Fig. 5-4(c). This also represents, to a different scale, the flux-density distribution. The mmf distribution of Fig. 5-4(c) can be Fourier analyzed, and the fundamental component is given by

$$\mathscr{F}_1(x,\ t)\ =\ \frac{4}{\pi}\left(\frac{Ni}{2}\right)\cos\frac{\pi x}{\tau} \qquad (5.1)$$

where x measures circumferential distance around the stator and where τ, the pole pitch (or coil pitch), is the circumferential distance between adjacent poles. If $i = N\sqrt{2}\ \sin \omega t$ (A), (5.1) becomes

$$\mathscr{F}_1(x,\ t)\ =\ 0.9 NI\ \sin \omega t\ \cos\frac{\pi x}{\tau}\quad \text{(At)} \qquad (5.2)$$

I being the rms value of i.

To eliminate the harmonics from the mmfs, the armature winding is appropriately distributed over the stator periphery, as in Fig. 5-3. Thus, we may assume that the mmf produced by each phase (of a three-phase winding, say) is sinusoidal in space. In a three-phase induction machine the mmfs are displaced from each other by 120° (electrical) in space:

$$\mathscr{F}_A\ =\ \mathscr{F}_m\ \sin \omega t\ \cos\frac{\pi x}{\tau}$$

$$\mathscr{F}_B\ =\ \mathscr{F}_m\ \sin (\omega t\ -\ 120°)\ \cos\left(\frac{\pi x}{\tau}\ -\ 120°\right) \qquad (5.3)$$

$$\mathscr{F}_C\ =\ \mathscr{F}_m\ \sin (\omega t\ +\ 120°)\ \cos\left(\frac{\pi x}{\tau}\ +\ 120°\right)$$

where $\mathscr{F}_m$ is the amplitude of each mmf. For the N-turn coil, considering only the fundamental, $\mathscr{F}_m = 0.9 NI$.

5.3 PRODUCTION OF ROTATING MAGNETIC FIELDS

Adding the three mmfs of (5.3), we obtain the resultant mmf as

$$\mathscr{F}(x,\ t)\ =\ 1.5\mathscr{F}_m\ \sin\left(\omega t\ -\ \frac{\pi x}{\tau}\right) \qquad (5.4)$$

It is seen that the mmf is a wave, of amplitude $1.5\mathscr{F}_m$, that travels circumferentially at speed

$$v_s\ \equiv\ \frac{\tau\omega}{\pi}\quad \text{(m/s)} \qquad (5.5)$$

relative to the stator. We call v_s the *synchronous velocity*. Note that the wavelength is

$$\lambda\ =\ \frac{2\pi v_s}{\omega}\ =\ 2\tau\quad \text{(m)} \qquad (5.6)$$

If the machine has p poles, (5.5) may be rewritten in the form

$$n_s\ \equiv\ \text{synchronous speed}\ =\ \frac{120 f_1}{p}\quad \text{(rpm)} \qquad (5.7)$$

where $f_1 = \omega/2\pi$ is the stator current (and mmf-rotational) frequency.

Equation (5.4) describes the rotating magnetic field produced by the stator of the induction motor. This field cuts the rotor conductors, and thereby voltages are induced in these conductors. The induced voltages give rise to rotor currents, which interact with the airgap field to produce a torque, which is maintained as long as the rotating magnetic field and induced rotor currents exist. Consequently, the motor starts rotating at a speed $n < n_s$ in the direction of the rotating field. (See Problem 5.26.)

5.4 SLIP; MACHINE EQUIVALENT CIRCUITS

The actual speed, n, of the rotor is often related to the synchronous speed, n_s, via the *slip*:

$$s \equiv \frac{n_s - n}{n_s} \qquad (5.8)$$

or the *percent slip*, $100s$.

At standstill ($s = 1$), the rotating magnetic field produced by the stator has the same speed with respect to the rotor windings as with respect to the stator windings. Thus, the frequency of the rotor currents, f_2, is the same as the frequency of the stator currents, f_1. At synchronous speed ($s = 0$), there is no relative motion between the rotating field and the rotor, and the frequency of rotor current is zero. (Indeed, the rotor current is zero.) At intermediate speeds the rotor current frequency is proportional to the slip,

$$f_s = sf_1 \qquad (5.9)$$

where f_2 is known as the *slip frequency*. Noting that the rotor currents are of slip frequency, we have the rotor equivalent circuit (on a per-phase basis) of Fig. 5-5(a), which gives the rotor current, I_2, as

$$I_2 = \frac{sE_2}{\sqrt{R_2^2 + (sX_2)^2}}$$

Here, E_2 is the induced rotor emf at standstill; X_2 is the rotor leakage reactance per phase at standstill; and R_2 is the rotor resistance per phase. This may also be written as

$$I_2 = \frac{E_2}{\sqrt{(R_2/s)^2 + X_2^2}} \qquad (5.10)$$

For (5.10) we redraw the circuit of Fig. 5-5(a) as Fig. 5-5(b).

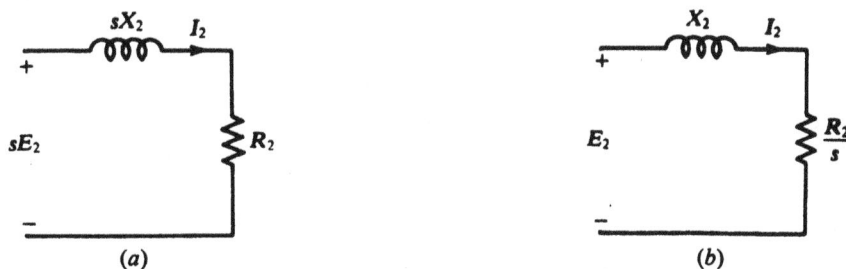

$$(a) \qquad\qquad\qquad\qquad (b)$$

Fig. 5-5

In order to include the stator circuit, the induction motor may be viewed as a transformer with an airgap, having a variable resistance in the secondary [see (5.10)]. Thus the primary of the transformer corresponds to the stator of the induction motor, whereas the secondary corresponds to the rotor on a per-phase basis. Because of the airgap, however, the value of the magnetizing reactance, X_m, tends to be low as compared to that of a true transformer. As in a transformer, we have a mutual flux linking both the stator and rotor, represented by the magnetizing reactance and various leakage fluxes. For instance, the total rotor leakage flux is denoted by X_2 in Fig. 5-5(b). Considering the rotor as being coupled to the stator as the secondary of a transformer is coupled to its primary, we may draw the circuit shown in Fig. 5-6. To develop this circuit further, we need to express the rotor quantities as referred to the stator. For this purpose we must know the transformation ratio, as in a transformer.

The voltage transformation ratio in the induction motor must include the effect of the stator and rotor winding distributions. It can be shown that, for a cage-type rotor, the rotor resistance per phase, R_2' referred

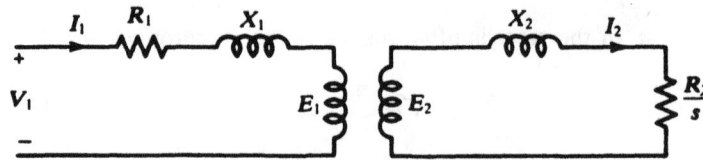

Fig. 5-6

to the stator, is

$$R_2' = a^2 R_2 \quad \text{where} \quad a^2 \equiv \frac{m_1}{m_2}\left(\frac{k_{w1}N_1}{k_{w2}N_2}\right)^2$$

Here $k_{w1} \equiv$ winding factor (see Problem 5.3) of the stator having N_1 series-connected turns per phase
$\quad\quad k_{w2} \equiv$ winding factor of the rotor having $N_2 = p/4$ series-connected turns per phase, for a cage rotor,
$\quad\quad\quad$ where p is the number of poles
$\quad\quad m_1 \equiv$ number of phase on the stator
$\quad\quad m_2 \equiv$ number of bars per pole pair
$\quad\quad R_2 \equiv$ resistance of one bar

Similarly,

$$X_2' = a^2 X_2 \tag{5.12}$$

where X_2' is the rotor leakage reactance per phase, referred to the stator.

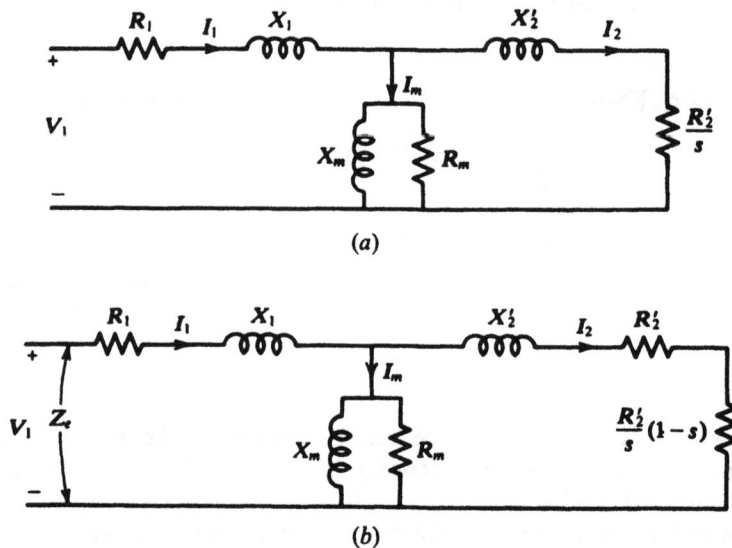

(a)

(b)

Fig. 5-7

Bearing in mind both the similarities and the differences between an induction motor and a transformer, we now refer the rotor quantities to the stator to obtain from Fig. 5-6 the exact equivalent circuit (per phase) shown in Fig. 5-7(a). For reasons that will become immediately clear, we split R_2'/s as

$$\frac{R_2'}{s} \equiv R_2' + \frac{R_2'}{s}(1-s)$$

to obtain the circuit shown in Fig. 5-7(b). Here, R_2' is simply the per-phase standstill rotor resistance referred to the stator and $R_2'(1-s)/s$ is a per-phase dynamic resistance that depends on the rotor speed and corresponds to the load on the motor. Notice that all the parameters shown in Fig. 5-7 are standstill values.

5.5 CALCULATIONS FROM EQUIVALENT CIRCUITS

The major usefulness of an equivalent circuit of an induction motor is in the calculation of its performance. All calculations are made on a per-phase basis, assuming a balanced operation of the machine; the total quantities are then obtained by using the appropriate multiplying factor.

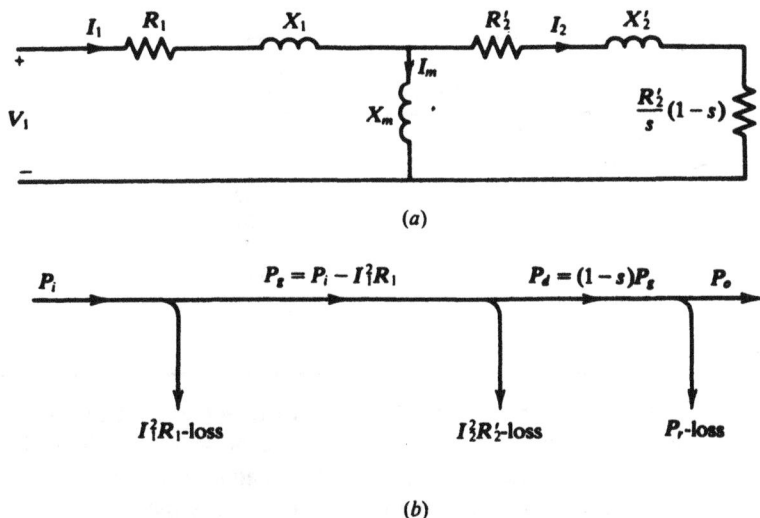

(a)

(b)

Fig. 5-8. Power flow in an induction motor.

Figure 5-8(a) is Fig. 5-7(b) with R_m omitted. (Core losses, most of which are in the stator, will be included only in efficiency calculations.) In Fig. 5-8(b) we show approximately the power flow and various power losses in one phase of the machine. The power crossing the airgap, P_g, is the difference between the input power, $P_i = V_1 I_1 \cos\theta_1$, and the stator resistive loss; that is,

$$P_g = P_i - I_1^2 R_1 \tag{5.13}$$

This power is dissipated in the net resistance R_2'/s, whence

$$P_g = I_2^2 \frac{R_2'}{s} \tag{5.14}$$

If we subtract the rotor (standstill) resistive loss from P_g, we obtain the developed electromagnetic power, P_d, so that

$$P_d = P_g - I_2^2 R_2' = (1-s)P_g \tag{5.15}$$

This is the power that appears across the resistance $R_2'(1-s)/s$, which corresponds to the load. The rotational (mechanical) loss, P_r, may be subtracted from P_d to obtain the shaft output power, P_o. Thus

$$P_o = P_d - P_r \tag{5.16}$$

and the efficiency, η, is the ratio P_o/P_i.

5.8 ENERGY-EFFICIENT INDUCTION MOTORS

It has been reported that the annual energy cost to operate a 10-hp induction motor 4000 h per year has increased from $850 in 1972 to $1950 in 1980. The escalation of oil prices in the mid-1970s led the manufacturers of electric motors to seek methods to improve motor efficiencies. In order to improve the motor efficiency, its loss distribution must be studied. For a typical standard three-phase 50-hp motor, the loss distribution at full-load is given in Table 5-1. In this table we also show the average loss distribution in percent of total losses for standard induction motors. The per unit loss in Table 5.1 is defined as loss/(hp × 746).

Table 5-1. Loss distribution in standard induction motors

Loss Distribution	50-hp Motor			Average Percent Loss for Standard Motors
	Watts	Percent Loss	Per Unit Loss	
Stator I^2R loss	1,540	38	0.04	37
Rotor I^2R loss	860	22	0.02	18
Magnetic core loss	765	20	0.02	20
Friction and windage loss	300	8	0.01	9
Stray load loss	452	12	0.01	16
Total losses	3,917	100	0.10	
Output (W)	37,300			
Input (W)	41,217			
Efficiency (%)	90.5			

In improving the efficiency of the motor, we must design to achieve a balance among the various losses and, at the same time, meet other specifications, such as breakdown torque, locked-rotor current and torque, and power factor. For the motor designer, a clear understanding of the loss distribution is very important. Loss reductions can be made by increasing the amount of the material in the motor. Without making other major design changes, a loss reduction of about 10 percent at full load can be achieved. Improving the magnetic circuit design using lower-loss electrical grade laminations can result in a further reduction of losses by about 10 percent. The cost of improving the motor efficiency increases with output rating (hp) of the motor. Based on the improvements just mentioned to increase the motor efficiency, Fig. 5.9 shows a comparison between the efficiencies of energy-efficient motors and those of standard motors.

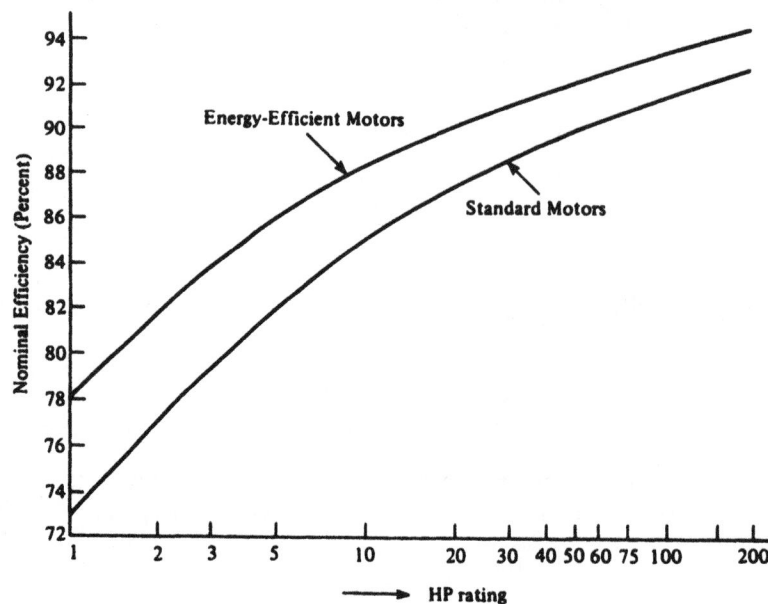

Fig. 5-9

Several of the major manufacturers of induction motors have developed product lines of energy-efficient motors. These motors are identified by their trade names, such as:

E-Plus (Gould Inc.)
Energy Saver (General Electric)
XE-Energy Efficient (Reliance Electric)
Mac II High Efficiency (Westinghouse)

Because energy-efficient motors use more material, they are relatively bigger in size compared to standard motors.

Example

A major manufacturer of home and office air conditioners uses a ½-hp, single-phase induction motor that has an efficiency of 72% at its average power output level. Several large-volume customers indicated that they would be willing to pay a larger initial investment in air conditioners if they could recover this increased investment during the warranty period, which is 2 years. The typical office air conditioner in which they are interested runs approximately 8 hours per day during 140 equivalent days of an Atlanta year. The motor supplied to the manufacturer has a wholesale cost of $45. If the motor supplier would achieve an average efficiency of 85% by improving materials and design, how much cost differential could be added to the wholesale motor cost and still satisfy the customer's request?

$$\text{Energy used by one present motor in 2 years} = \frac{1}{2} \times \frac{746}{0.72} \times \frac{140 \times 8 \times 2}{1000} = 1160 \text{ kWh}$$

$$\text{At 0.85 efficiency, the consumed energy} = \frac{1}{2} \times \frac{746}{0.85} \times \frac{140 \times 8 \times 2}{1000} = 983 \text{ kWh}$$

Energy savings = 1160 − 83 = 177 kWh
At an energy cost of 7¢/kWh, energy savings in dollars = 177 × 0.07 = $12.39
Thus $12.39 could be added to the ($45) initial cost of the motor.

5.7 APPROXIMATE EQUIVALENT CIRCUIT PARAMETERS FROM TEST DATA

Sometimes the equivalent circuit of the induction motor is approximated by the one shown in Fig. 5-10. The parameters of the approximate circuit can be obtained from the following two tests.

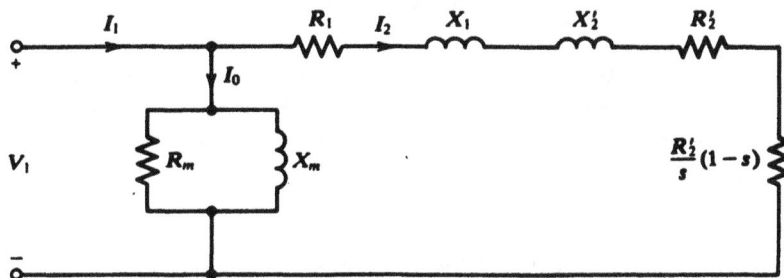

Fig. 5-10

No-Load Test

In this test, rated voltage is applied to the machine and it is allowed to run on no-load. Input power (corrected for friction and windage loss), voltage, and current are measured; these, reduced to per-phase values, are denoted by P_0, V_0, and I_0, respectively. When the machine runs on no-load, the slip is close to zero and the circuit in Fig. 5-10 to the right of the shunt branch is taken to be an open circuit. Thus, the parameters R_m and X_m are found from

$$R_m = \frac{V_0^2}{P_0} \qquad (5.17)$$

$$X_m = \frac{V_0^2}{\sqrt{V_0^2 I_0^2 - P_0^2}} \qquad (5.18)$$

Blocked-Rotor Test

In this test, the rotor of the machine is blocked ($s = 1$), and a reduced voltage is applied to the machine so that the rated current flows through the stator windings. The input power, voltage, and current are recorded and reduced to per-phase values; these are denoted, respectively, by P_s, V_s, and I_s. In this test, the iron losses are assumed to be negligible and the shunt branch of the circuit shown in Fig. 5-10 is considered to be absent. The parameters are thus found from

$$R_e = R_1 + a^2 R_2 = \frac{P_s}{I_s^2} \qquad (5.19)$$

$$X_e = X_1 + a^2 X_2 = \frac{\sqrt{V_s^2 I_s^2 - P_s^2}}{I_s^2} \qquad (5.20)$$

In (5.19) and (5.20), the constant a^2 is the same as in (5.11). The stator resistance per phase, R_1, can be directly measured, and, knowing R_e from (5.19), we can determine $R_2' = a^2 R_2$, the rotor resistance referred to the stator. There is no simple method of determining X_1 and $X_2' = a^2 X_2$ separately. The total value given by (5.20) is sometimes equally divided between X_1 and X_2'.

Solved Problems

5.1. An N-turn winding is made up of coils distributed in slots, as the winding shown in Fig. 5-3. The voltages induced in these coils are displaced from one another in phase by the slot angle α. The resultant voltage at the terminals of the N-turn winding is then the phasor sum of the coil voltages. Find an expression for the *distribution factor*, k_d, where

$$k_d \equiv \frac{\text{magnitude of resultant voltage}}{\text{sum of magnitudes of individual coil voltages}}$$

Let p be the number of poles; Q, the number of slots; and m, the number of phases. Then $Q = qpm$, where q is the number of slots per pole per phase. The slot angle α is given (in electrical degrees) by

$$\alpha = \frac{(180°)p}{Q} = \frac{180°}{mq}$$

The phasor addition of voltages (for $q = 3$) is shown in Fig. 5-11, from the geometry of which we get

$$k_d = \frac{E_r}{qE_c} = \frac{2l \sin(q\alpha/2)}{q[2l \sin(\alpha/2)]} = \frac{\sin(q\alpha/2)}{q \sin(\alpha/2)} \qquad (5.21)$$

which is the desired result.

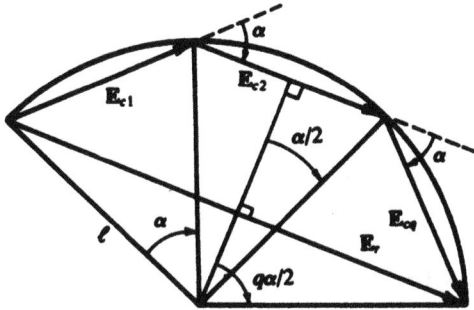

Fig. 5-11

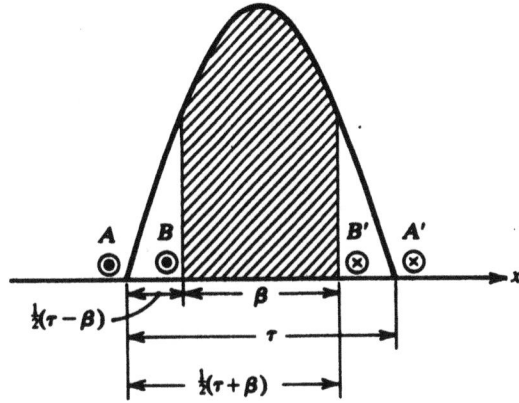

Fig. 5-12

5.2. The voltage induced in a fractional-pitch coil is reduced by a factor known as the *pitch factor*, k_p, as compared to the voltage induced in a full-pitch coil. Derive an expression for the pitch factor.

In a sinusoidally distributed flux density we show a full-pitch and a fractional-pitch coil in Fig. 5-12. The coil span of the full-pitch coil is equal to the pole pitch, τ. Let the coil span of the fractional-pitch coil be $\beta < \tau$, as shown. The flux linking the fractional-pitch coil will be proportional to the shaded area in Fig. 5-12 whereas the flux linking the full-pitch coil is proportional to the entire area under the curve. The pitch factor is therefore the ratio of the shaded area to the total area:

$$k_p = \frac{\int_{(\tau-\beta)/2}^{(\tau+\beta)/2} \sin \frac{\pi x}{\tau} \, dx}{\int_0^\tau \sin \frac{\pi x}{\tau} \, dx} = \sin \frac{\pi\beta}{2\tau} \qquad (5.22)$$

Notice that in (5.22), β and τ may be measured in any convenient unit.

5.3. Calculate the distribution factor (Problem 5.1), the pitch factor (Problem 5.2), and the *winding factor*, $k_w \equiv k_d k_p$, for the stator winding of Fig. 5-3.

From Fig. 5-3, $m = 3$, $p = 4$, and $Q = 36$. Thus,

$$q = \frac{36}{(4)(3)} = 3 \qquad \alpha = \frac{180°}{(3)(3)} = 20°$$

Substituting these in (5.21) yields

$$k_d = \frac{\sin 30°}{3 \sin 10°} = 0.96$$

Also, Fig. 5-3 shows that $\tau = 9$ slots and $\beta = 8$ slots. Hence, from (5.22),

$$k_p = \sin \frac{8\pi}{18} = \sin 80° = 0.985$$

and

$$k_w = k_d k_p = (0.96)(0.985) = 0.945$$

5.4. A 4-pole, 3-phase induction motor is energized from a 60-Hz supply, and is running at a load condition for which the slip is 0.03. Determine: (a) rotor speed, in rpm; (b) rotor current frequency, in Hz; (c) speed of the rotor rotating magnetic field with respect to the stator frame, in rpm; (d) speed of the rotor rotating magnetic field with respect to the stator rotating magnetic field, in rpm.

$$n_s = \frac{120f_1}{p} = \frac{120(60)}{4} = 1800 \text{ rpm}$$

(a)
$$n = (1 - s)n_s = (1 - 0.03)(1800) = 1746 \text{ rpm}$$

(b)
$$f_2 = sf_1 = (0.03)(60) = 1.8 \text{ Hz}$$

(c) The p poles on the stator induce an equal number of poles on the rotor. Now, the same argument that led to (5.4) can be applied to the rotor. Thus, the rotor produces a rotating magnetic field whose speed, *relative to the rotor*, is

$$n_r = \frac{120f_2}{p} = \frac{120sf_1}{p} = sn_s$$

But the speed of the rotor relative to the stator is $n = (1 - s)n_s$. Therefore, the speed of the rotor field with respect to the stator is

$$n_s' = n_r + n = n_s$$

i.e., in this case, 1800 rpm.

(d) Zero.

5.5. A 60-Hz induction motor has 2 poles and runs at 3510 rpm. Calculate (a) the synchronous speed and (b) the percent slip.

(a)
$$n_s = \frac{120f_1}{p} = \frac{120(60)}{2} = 3600 \text{ rpm}$$

(b)
$$s = \frac{n_s - n}{n_s} = \frac{3600 - 3510}{3600} = 0.025 = 2.5\%$$

5.6. Using the rotor equivalent circuit of Fig. 5-5(b), show that an induction motor will have a maximum starting torque when its rotor resistance (regarded as variable) is equal to its leakage reactance. All quantities are on a per-phase basis.

From Fig. 5-5(b), the developed power, P_d, is given by

$$P_d = I_2^2 \frac{R_2}{s} - I_2^2 R_2 = T_e \omega_m \tag{1}$$

and the rotor current I_2 is such that

$$I_2^2 = \frac{E_2^2}{(R_2/s)^2 + X_2^2} \tag{2}$$

Also, the mechanical angular velocity is

$$\omega_m = (1 - s)\omega_s \tag{3}$$

where ω_s is the synchronous angular velocity. These three equations give:

$$T_e = \frac{E_2^2 s}{\omega_s} \frac{R_2}{R_2^2 + s^2 X_2^2} \tag{4}$$

For a maximum T_e we must have $\partial T_e/\partial R_2 = 0$, which leads to

$$R_2^2 + s^2 X_2^2 - 2R_2^2 = 0 \quad \text{or} \quad R_2 = sX_2$$

At starting, $s = 1$, this becomes $R_2 = X_2$.

5.7. Using only the rotor circuit (as in Problem 5.6), calculate the torque developed per phase by a 6-pole, 60-Hz, 3-phase induction motor at a slip of 5%, if the motor develops a maximum per-phase torque $T_e^* = 300$ N · m while running at 780 rpm. The rotor leakage reactance is 3.0 Ω per phase.

$$n_s = \frac{120(60)}{6} = 1200 \text{ rpm}$$

At 780 rpm,

$$s^* = \frac{1200 - 780}{1200} = 0.35$$

so that $s^*/s = 0.35/0.05 = 7$. From (4) of Problem 5.6,

$$\frac{T_e}{T_e^*} = \frac{s}{s^*} \frac{R_2^2 + s^{*2} X_2^2}{R_2^2 + s^2 X_2^2} = \frac{2(s^*/s)}{1 + (s^*/s)^2}$$

where we have used $R_2 = s^* X_2$. Consequently,

$$T_e = \frac{2(7)}{1 + (7)^2} (300) = 84 \text{ N · m}$$

5.8. The rotor of a 3-phase, 60-Hz, 4-pole induction motor takes 120 kW at 3 Hz. Determine (a) the rotor speed and (b) the rotor copper losses.

(a)　　$s = \frac{f_2}{f_1} = \frac{3}{60} + 0.05 \qquad n_s = \frac{120 f_1}{p} = \frac{120(60)}{4} = 1800 \text{ rpm}$

$$n = (1 - s)n_s = (1 - 0.05)(1800) = 1710 \text{ rpm}$$

(b)   By (5.15),

$$\text{rotor copper loss} = s \times (\text{rotor input}) = (0.05)(120) = 6 \text{ kW}$$

5.9. The motor of Problem 5.8 has a stator copper loss of 3 kW, a mechanical loss of 2 kW, and a stator core loss of 1.7 kW. Calculate (a) the motor output at the shaft and (b) the efficiency. Neglect rotor core loss.

From Problem 5.8, the rotor input is 120 kW and the rotor copper loss is 6 kW.

(a) $\text{motor output} = 120 - 6 - 2 = 112 \text{ kW}$

(b) $\text{motor input} = 120 + 3 + 1.7 = 124.7 \text{ kW}$

$$\text{efficiency} = \frac{\text{output}}{\text{input}} = \frac{112}{124.7} = 89.8\%$$

5.10. A 6-pole, 3-phase, 60-Hz induction motor takes 48 kW in power at 1140 rpm. The stator copper loss is 1.4 kW, stator core loss is 1.6 kW, and rotor mechanical losses are 1 kW. Find the motor efficiency.

$$n_s = \frac{120 f_1}{p} = \frac{120(60)}{6} = 1200 \text{ rpm} \qquad s = \frac{n_s - n}{n_s} = \frac{1200 - 1140}{1200} = 0.05$$

$\text{rotor input} = \text{stator output} = (\text{stator input}) - (\text{stator losses}) = 48 - (1.4 + 1.6) = 45 \text{ kW}$

$\text{rotor output} = (1 - s) \times (\text{rotor input}) = (1 - 0.05)(45) = 42.75 \text{ kW}$

$\text{motor output} = (\text{rotor output}) - (\text{rotational losses}) = 42.75 - 1 = 41.75 \text{ kW}$

$\text{motor efficiency} = \frac{41.75}{48} = 87\%$

5.11. A slip-ring induction motor, having a synchronous speed of 1800 rpm, runs at $n = 1710$ rpm when the rotor resistance per phase is 0.2 Ω. The motor is required to develop a constant torque down to a speed of $n^* = 1440$ rpm. Using the rotor circuit of Fig. 5-5(b), explain how this goal may be accomplished. The rotor leakage reactance at standstill is 2 Ω per phase.

From (4) of Problem 5.6, we may write

$$T_e = k \frac{s R_2}{R_2^2 + s^2 X_2^2}$$

where $k \equiv E_2^2/\omega s$ is a positive constant. It is easy to verify that $\partial T_e/\partial s$ and $\partial T_e/\partial R_2$ are always of opposite signs. Thus, if T_e is to stay fixed as s increases (i.e., the speed decreases), R_2 must also continuously increase, attaining its maximum value of s^*. We then have the quadratic equation

$$\frac{s R_2}{R_2^2 + s^2 X_2^2} = \frac{s^* R_2^*}{R_2^{*2} + s^{*2} X_2^2}$$

for this maximum value, R_2^*. Substituting the numerical data

$$s = \frac{1800 - 1710}{1800} = 0.05 \qquad s^* = \frac{1800 - 1440}{1800} = 0.2$$

$$R_2 = 0.2\ \Omega \qquad X_2 = 2\ \Omega$$

and solving, we obtain $R_2^* = 0.8\ \Omega$. Thus, a continuously variable external resistor, of maximum resistance $0.8 - 0.2 = 0.6\ \Omega$, must be inserted in the rotor circuit.

5.12. The synchronous speed of an induction motor is 900 rpm. Under a blocked-rotor condition, the input power to the motor is 45 kW at 193.6 A. The stator resistance per phase is 0.2 Ω and the transformation ratio is $a = 2$. Calculate (a) the ohmic value of the rotor resistance per phase and (b) the motor starting torque. The stator and rotor are wye-connected.

(a)   From (5.19)

$$R_1 + a^2 R_2 = \frac{P_s}{I_s^2} \quad \text{or} \quad 0.2 + 4R_2 = \frac{(45 \times 10^3)/3}{(193.6)^2}$$

whence $R_2 = 0.05\ \Omega$.

(b)   Referred to the stator, the rotor resistance per phase is $R_2' = a^2 R_2 = 0.2\ \Omega$. Then

$$\text{starting torque} = \frac{3I_s^2 R_2'}{\omega_s} = \frac{3(193.6)^2(0.2)}{2\pi(900)/60} = 238.6\ \text{N} \cdot \text{m}$$

5.13. A 3-phase induction motor has the per-phase circuit parameters shown in Fig. 5-13. At what slip will the developed power be maximum?

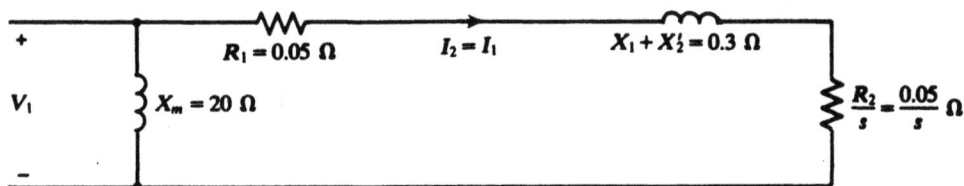

Fig. 5-13

The developed power per phase is given by

$$P_d = I_2^2 \frac{R_2}{s} (1 - s) \tag{1}$$

From the circuit,

$$I_2^2 = \frac{V_1^2}{(R_1 + R_2/s)^2 + (X_1 + X_2')^2}$$

Substituting this in (1) and inserting numerical values yields

$$P_d = \text{constant} \times \frac{s(1 - s)}{(s + 1)^2 + 36s^2}$$

Setting $\partial P_d/\partial s = 0$, we obtain a quadratic equation for s, the solution of which is $s \approx 0.14$.

5.14. The per-phase parameters of the equivalent circuit, Fig. 5-8(a), for a 400-V, 60-Hz, 3-phase, wye-connected, 4-pole induction motor are:

$$R_1 = 2R_2' = 0.2 \ \Omega \qquad X_1 = 0.5 \ \Omega \qquad X_2' = 0.2 \ \Omega \qquad X_m = 20 \ \Omega$$

If the total mechanical and iron losses at 1755 rpm are 800 W, compute (a) input current, (b) input power, (c) output power, (d) output torque, and (e) efficiency (all at 1755 rpm).

$$n_s = \frac{120(60)}{4} = 1800 \text{ rpm} \qquad s = \frac{1800 - 1755}{1800} = \frac{1}{40}$$

From the given circuit, the equivalent impedance per phase is

$$\mathbf{Z}_e = (0.2 + j0.5) + \frac{(j20)(4 + j0.2)}{4 + j(20 + 0.2)}$$

$$= (0.2 + j0.5) + (3.77 + j0.944) = 4.223\angle20° \ \Omega$$

and the phase voltage is $400/\sqrt{3} = 231$ V.

(a)
$$\text{input current} = \frac{231}{4.223} = 54.65 \text{ A}$$

(b)
$$\text{total input power} = \sqrt{3} \ (400)(54.65)(\cos 20°) = 35.58 \text{ kW}$$

(c) The total power crossing the airgap, P_g, is the power in the three 3.77 Ω resistances (see the expression for $\mathbf{Z}_e$ above). Thus,

$$P_g = 3(54.65)^2(3.77) = 33.789 \text{ kW}$$

[Or, by subtraction of the stator losses, $P_g = 35\ 580 - 3(54.65)^2(0.2) = 33.788$ kW.] The total developed power is then

$$P_d = (1 - s)P_g = (0.975)(33.79) = 32.94 \text{ kW}$$

and the total output power is

$$P_o = P_d - (800 \text{ W}) = 32.14 \text{ kW}$$

(d)
$$\text{output torque} = \frac{P_o}{\omega_m} = \frac{32140}{2\pi(1755)/60} = 174.9 \text{ N} \cdot \text{m}$$

(e)
$$\text{efficiency} = \frac{32.14}{35.58} = 90.3\%$$

5.15. The results of no-load and blocked-rotor tests on a 3-phase, wye-connected induction motor are as follows:

    no-load test: line-to-line voltage = 400 V
                               input power = 1770 W
                               input current = 18.5 A
                               friction and windage loss = 600 W

blocked-rotor test: line-to-line voltage = 45 V
input power = 2700 W
input current = 63 A

Determine the parameters of the approximate equivalent circuit (Fig. 5-10).

From no-load test data:

$$V_0 = \frac{400}{\sqrt{3}} = 231 \text{ V} \qquad P_0 = \frac{1}{3}(1770 - 600) = 390 \text{ W} \qquad I_0 = 18.5 \text{ A}$$

Then, by (5.17) and (5.18),

$$R_m = \frac{(231)^2}{390} = 136.8 \ \Omega$$

$$X_m = \frac{(231)^2}{\sqrt{(231)^2(18.5)^2 - (390)^2}} = 12.5 \ \Omega$$

From blocked-rotor test data:

$$V_s = \frac{45}{\sqrt{3}} = 25.98 \text{ V} \qquad I_s = 63 \text{ A} \qquad P_s = \frac{2700}{3} = 900 \text{ W}$$

Then, by (5.19) and (5.20),

$$R_e = R_1 + a^2R_2 = \frac{900}{(63)^2} = 0.23 \ \Omega$$

$$X_e = X_1 + a^2X_2 = \frac{\sqrt{(25.98)^2(63)^2 - (900)^2}}{(63)^2} = 0.34 \ \Omega$$

5.16   (a) Replace the circuit of Fig. 5-8(a) by its Thevenin equivalent circuit and express the Thevenin voltage, V_{Th}, and impedance, $Z_{Th} = R_{Th} + jX_{Th}$, in terms of the circuit parameters of Fig. 5-8(a) and the voltage V_1. (b) The per-phase parameters for Fig. 5-8(a) are as in Problem 5.14. Other data also remain the same. Draw a Thevenin equivalent circuit for the motor.

(a)   From Fig. 5-8(a),

$$V_{Th} = \frac{jX_m}{R_1 + j(X_1 + X_m)} V_1 \qquad Z_{Th} = \frac{jX_m(R_1 + jX_1)}{R_1 + j(X_1 + X_m)}$$

(b)   The Thevenin circuit is shown in Fig. 5-14, for which the numerical values are:

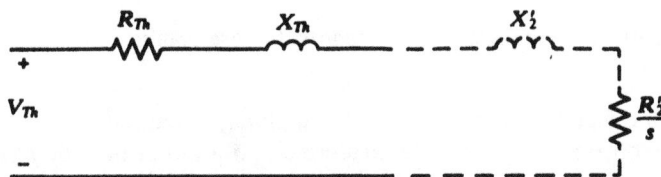

Fig. 5-14

$$V_{Th} = \frac{400}{\sqrt{3}} \frac{j20}{0.2 + j20.5} \quad \text{or} \quad V_{Th} = 225.3 \text{ V}$$

$$R_{Th} + jX_{Th} = \frac{j20(0.2 + j0.5)}{0.2 + j20.5} = 0.19 + j0.49 \ \Omega$$

5.17. Compute the starting current and starting torque of the motor of Problem 5.16.

Use the complete circuit of Fig. 5-14, with [see Problem 5.16(b) and Problem 5.14] $V_{Th} = 225.3$ V, $R_{Th} = 0.19 \ \Omega$, $X_{Th} = 0.49 \ \Omega$, $X_2' = 0.2 \ \Omega$, $R_2'/s = R_2' = 0.1 \ \Omega$. Thus,

$$I_{s=1} = \frac{225.3}{[(0.19 + 0.1)^2 + (0.49 + 0.2)^2]^{1/2}} = 301 \text{ A}$$

The starting torque is given by the expression developed in Problem 5.12(b):

$$T_{s=1} = \frac{3(301)^2(0.1)}{2\pi(1800)/60} = 144.2 \text{ N} \cdot \text{m}$$

5.18 For the data of Problem 5.14, using the complete circuit of Fig. 5-14, calculate (a) power crossing the airgap, (b) developed power, (c) output power, and (d) output torque of the motor. Compare with the corresponding results of Problem 5.14.

From Fig. 5-14, with $s = 1/40$,

$$\mathbf{Z}_e = \mathbf{Z}_{Th} + \frac{R_2'}{s} + jX_2' = 0.19 + j0.49 + 4 + j0.2$$

whence, $\mathbf{Z}_e = 4.246 \ \Omega$. Moreover, as computed in Problem 5.16(b), $V_{Th} = 225.3$ V. Then,

$$I_2' = \frac{V_{Th}}{\mathbf{Z}_e} = \frac{225.3}{4.246} = 53.06 \text{ A}$$

(a)
$$P_g = 3I_2'^2 \frac{R_2'}{s} = 3(53.06)^2(4) = 33.784 \text{ kW}$$

(b)
$$P_d = (1 - s)P_g = (0.975)(33.784) = 32.939 \text{ kW}$$

(c)
$$P_o = P_d - (800 \text{ W}) = 32.139 \text{ kW}$$

(d)
$$T_o = \frac{P_o}{\omega_m} = \frac{32\,139}{2\pi(1755)/60} = 174.9 \text{ N} \cdot \text{m}$$

The results are in excellent agreement with those obtained in Problem 5.14.

5.19. A large induction motor is usually started by applying a reduced voltage across the motor; such a voltage may be obtained from an autotransformer. If a motor is to be started on 50% of full-load torque and if the full-voltage starting current is 5 times the full-load current, determine the percent reduction in the applied voltage (that is, the percent tap on the autotransformer). The full-load slip is 4%.

Recall that

$$T = \frac{I_2'^2 R_2'}{s\omega_s} \quad \text{where } I_2' \approx I_1 \tag{1}$$

Let us write:

$(I_2')_{\text{SFV}} \equiv$ rotor current at start if full voltage is applied

$(I_2')_S \equiv$ rotor current at start if reduced voltage is applied

$k \equiv$ ratio of reduced voltage to full voltage

At a given slip—in particular, at $s = 1$—rotor current may be considered proportional to applied voltage. Hence,

$$\frac{(I_2')_S}{(I_2')_{\text{SFV}}} = k$$

and

$$(I_2')_S = k(I_2')_{\text{SFV}} \approx k5I_{2\text{FL}}' \tag{2}$$

Applying (1) at reduced-voltage start and at full-load, and substituting (2), we obtain

$$\frac{T_S}{T_{\text{FL}}} \approx \left(\frac{k5I_{2\text{FL}}'}{I_{2\text{FL}}'}\right)\frac{s_{\text{FL}}}{1} \quad \text{or} \quad \frac{1}{2} \approx (25k^2)(0.04)$$

from which $k \approx 0.707 = 70.7\%$.

5.20. A 3-phase, 400-V, wye-connected induction motor takes the full-load current at 45 V with the rotor blocked. The full-load slip is 4%. Calculate the tappings k on a 3-phase autotransformer to limit the starting current to 4 times the full-load current. For such a limitation, determine the ratio of the starting torque to full-load torque.

By the (approximate) current-voltage proportionality,

$$\frac{I_b}{I_{\text{FL}}} = \frac{400}{45} \quad \text{and} \quad \frac{I_S}{I_b} = k$$

where I_b is the full-voltage blocked-rotor current, I_{FL} is the full-load current, and I_S is the starting current. But it is given that

$$\frac{I_S}{I_{\text{FL}}} = 4$$

Therefore,

$$k = \frac{I_S/I_{\text{FL}}}{I_b/I_{\text{FL}}} = \frac{4}{400/45} = 45\%$$

Now, from (1) of Problem 5.19,

$$\frac{T_S}{T_{FL}} = \left(\frac{I_S}{I_{FL}}\right)^2 s_{FL} = (4)^2(0.04) = 0.64$$

5.21. The motor of Problem 5.20 employs the *wye-delta starter* shown in Fig. 5-15; that is, the phases are connected in wye at the time of the starting and are switched to delta when the motor is running. The full-load slip is 4% and the motor draws approximately 9 times the full-load current if started directly from the mains. Determine the ratio of starting torque to full-load torque.

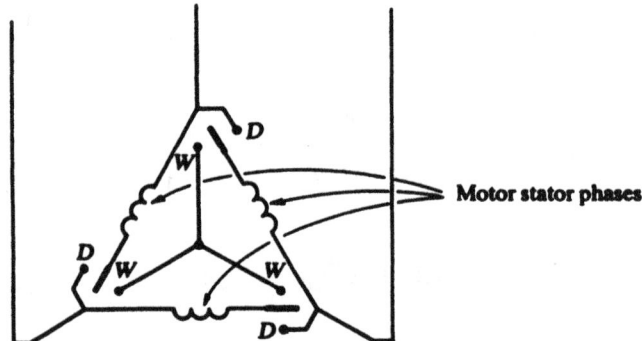

Fig. 5-15. Switches on *W* correspond to wye and switches on *D* correspond to the delta connection.

When the phases are switched to delta, the phase voltage, and hence the full-load current, is increased by a factor of $\sqrt{3}$ over the value it would have had in a wye connection. Then it follows from the last equation in Problem 5.20 that

$$\frac{T_S}{T_{FL}} = \left(\frac{9}{\sqrt{3}}\right)^2 (0.04) = 1.08$$

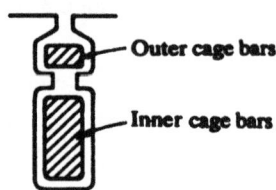

Fig. 5-16

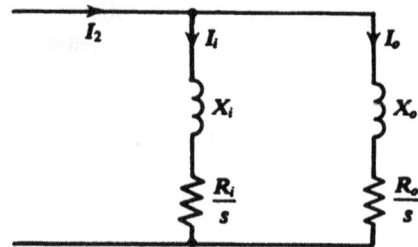

Fig. 5-17

5.22. To obtain a high starting torque in a cage-type motor, a double-cage rotor is used. The forms of a slot and of the bars of the two cages are shown in Fig. 5-16. The outer cage has a higher resistance than the inner cage. At starting, because of the skin-effect, the influence of the outer cage dominates, thus producing a high starting torque. An approximate equivalent circuit for such a rotor is given in Fig. 5-17. Suppose that, for a certain motor, we have the per-phase values

$$R_i = 0.1 \ \Omega \qquad R_0 = 1.2 \ \Omega \qquad X_i = 2 \ \Omega \qquad X_0 = 1 \ \Omega$$

Determine the ratio of the torques provided by the two cages at (*a*) starting and (*b*) 2% slip.

(a)   From Fig. 5-17, at $s = 1$,

$$Z_i^2 = (0.1)^2 + (2)^2 = 4.01 \ \Omega^2$$

$$Z_0^2 = (1.2)^2 + (1)^2 = 2.44 \ \Omega^2$$

power input to the inner cage $\equiv P_{ii} = I_i^2 R_i = 0.1 \ I_i^2$

power input to the outer cage $\equiv P_{io} = I_o^2 R_o = 1.2 \ I_o^2$

$$\frac{\text{torque due to inner cage}}{\text{torque due to outer cage}} \equiv \frac{T_i}{T_o} = \frac{P_{ii}}{P_{io}} = \frac{0.1}{1.2} \left(\frac{I_i}{I_o}\right)^2 = \frac{0.1}{1.2} \left(\frac{Z_o}{Z_i}\right)^2$$

$$= \frac{0.1}{1.2} \left(\frac{2.44}{4.01}\right) = 0.05$$

(b)   Similarly, at $s = 0.02$,

$$Z_i^2 = \left(\frac{0.1}{0.02}\right)^2 + (2)^2 = 29 \ \Omega^2$$

$$Z_o^2 = \left(\frac{1.2}{0.02}\right)^2 + (1)^2 = 3601 \ \Omega^2$$

$$\frac{T_i}{T_o} = \frac{0.1}{1.2} \left(\frac{3601}{29}\right) = 10.34$$

5.23.   At standstill, the impedances of the inner and outer cages of an induction motor are

$$\mathbf{Z}_i = 0.02 + j2 \ \Omega \qquad \mathbf{Z}_o = 0.2 + j1 \ \Omega$$

(per-phase values). At what slip will the torques contributed by the two cages be equal? Use the circuit of Fig. 5-17.

Let s be the required slip. Then,

$$Z_i^2 = \left(\frac{0.02}{s}\right)^2 + (2)^2 = \frac{4 \times 10^{-4}}{s^2} + 4$$

$$Z_o^2 = \left(\frac{0.2}{s}\right)^2 + (1)^2 = \frac{4 \times 10^{-2}}{s^2} + 1$$

$$P_i = I_i^2 \frac{R_i}{s} \qquad P_o = I_o^2 \frac{R_o}{s}$$

$$\frac{T_i}{T_o} = \left(\frac{I_i}{I_o}\right)^2 \frac{R_i}{R_o} = \left(\frac{Z_o}{Z_i}\right)^2 \frac{R_i}{R_o} = 1$$

which requires that

$$(0.02)\left(\frac{4 \times 10^{-2}}{s^2} + 1\right) = (0.2)\left(\frac{4 \times 10^{-4}}{s^2} + 4\right) \quad \text{or} \quad s = 0.03$$

5.24. At a slip of 3%, for the 3-phase motor of Problem 5.23, the rotor input phase voltage is 45 V. Calculate (a) the motor line current and (b) the torques contributed by the two cages. The motor has 4 poles and operates at 60 Hz.

At $s = 0.03$:

$$\mathbf{Z}_i = 0.67 + j2 \ \Omega \quad \mathbf{Z}_o = 6.67 + j1 \ \Omega$$

which are in parallel to form $\mathbf{Z}_e$. Hence,

$$\mathbf{Z}_e = \frac{(0.67 + j2)(6.67 + j1)}{(0.67 + j2) + (6.67 + j1)} = 0.95 + j1.5 = 1.8\angle 58° \ \Omega$$

(a)
$$I_2 = \frac{V_2}{Z_e} = \frac{45}{1.8} = 25 \text{ A}$$

$$n_s = \frac{120(60)}{4} = 1800 \text{ rpm}$$

(b)
$$\text{total torque} = \frac{3I_2^2 R_2}{s\omega_s} = \frac{3(25)^2(0.95)}{(0.03)(1800 \times 2\pi/60)} = 315 \text{ N} \cdot \text{m}$$

From Problem 5.23 we conclude that at $s = 0.03$ either cage contributes $315/2 = 157.5$ N · m.

Supplementary Problems

5.25. Verify the validity of (a) (5.4), (b) (5.7).

5.26. Explain why an induction motor will not run (a) at the synchronous speed, (b) in a direction opposite to the rotating magnetic field.

5.27. A 3-phase, distributed armature winding has 12 poles and 180 slots. The coil pitch is 14 slots. Calculate (a) the distribution factor, (b) the pitch factor, (c) the winding factor.
Ans. (a) 0.957; (b) 0.995; (c) 0.9517

5.28. A 3-phase, 60-Hz induction motor has 8 poles and operates with a slip of 0.05 for a certain load. Compute (in rpm) the (a) speed of the rotor with respect to the stator, (b) speed of the rotor with respect to the stator magnetic field, (c) speed of the rotor magnetic field with respect to the rotor, (d) speed of the rotor magnetic field with respect to the stator, (e) speed of the rotor field with respect to the stator field. *Ans.* (a) 855 rpm; (b) 45 rpm; (c) 45 rpm; (d) 990 rpm; (e) 0

5.29. A 3-phase, 60-Hz, 6-pole induction motor runs (a) on no-load at 1160 rpm and (b) on full-load at 1092 rpm. Determine the slip and frequency of rotor currents on no-load and on full-load.
Ans. (a) 0.034, 2 Hz; (b) 0.09, 5.4 Hz

5.30. An eight-pole induction motor is supplied from a 50-Hz source and runs at 720 rpm. Calculate the frequency of induced rotor current. *Ans.* 2 Hz

5.31. A 3-phase 4-pole wye-connected 440-V 60-Hz induction may be represented by the per phase approximate equivalent circuit shown in Fig. 5-13. The motor runs at 1710 rpm. Calculate (a) the developed electromagnetic torque and (b) the input power factor, if the per phase circuit parameters are: $X_1 = X_2'$ $= 0.25\ \Omega$; $R_1 = R_2 = 0.1\ \Omega$; and $X_m = 50\ \Omega$. *Ans.* (a) 450.6 N · m; (b) 0.962 lagging

5.32. Considering only the rotor circuit of the motor of Problem 5.31 determine the speed at which the motor will develop the maximum torque. *Ans.* 1080 rpm

5.33. The values of the approximate equivalent circuit parameters, Fig. 5-8(a), of a 3-phase, 600-V, 60-Hz, 4-pole, wye-connected induction motor are: $R_1 = 0.75\ \Omega$, $R_2' = 0.8\ \Omega$, $X_1 = X_2' = 2\ \Omega$, and $X_m = 50\ \Omega$. Obtain the values of the parameters and voltage of its Thevenin equivalent circuit.
Ans. $\mathbf{Z}_{Th} = 0.69 + j1.93\ \Omega$; $V_{Th} = 333$ V

5.34. From the Thevenin circuit, Fig. 5-14, show that (a) the slip s_m at which the maximum torque occurs is given by

$$s_m = \frac{R_2'}{\sqrt{R_{Th}^2 + (X_{Th} + X_2')^2}}$$

and (b) the corresponding maximum torque, T_m, can be expressed (on a per-phase basis) as

$$T_m = \frac{V_{Th}^2}{2\omega_s\left\{R_{Th} + [(X_{Th} + X_2')^2 + R_{Th}^2]^{1/2}\right\}}$$

5.35. Using Problem 5.34 and the data of Problem 5.33, calculate (a) the maximum torque (total) that the motor can develop, and (b) the corresponding speed. *Ans.* (a) 188 N · m; (b) 1440 rpm

5.36. A 3-phase, 60-Hz, 4-pole induction motor has a rotor leakage reactance of 0.8 Ω per phase and a rotor resistance of 0.1 Ω per phase. How much additional resistance must be inserted in the rotor circuit so that the motor shall have the maximum starting torque? Use the rotor circuit of Fig. 5-8(a) for your calculations. *Ans.* 0.7 Ω

5.37. A 20-hp, 3-phase, 400-V, 60-Hz, 4-pole induction motor delivers full-load at 5% slip. The mechanical rotational losses are 400 W. Calculate (a) the electromagnetic torque, (b) the shaft torque, and (c) the rotor copper loss. *Ans.* (a) 85.5 N · m; (b) 83.3 N · m; (c) 806.3 W

5.38. A 3-phase, 6-pole induction motor is rated 400 Hz, 150 V, 10 hp, 3% slip at rated power output. The windage and friction loss is 200 W at rated speed. With the motor operating at rated voltage, frequency, slip, and power output, determine (a) rotor speed, (b) frequency of rotor current, (c) rotor copper loss, (d) power crossing the airgap, (e) output torque.
Ans. (a) 7760 rpm; (b) 12 Hz; (c) 237 W; (d) 7897 W; (e) 9.2 N · m

5.39. The equivalent circuit and impedance values shown in Fig. 5-13 represent one phase of a 3-phase, wye-connected induction motor. With slip equal to 0.05 and with 100 V (line-to-neutral) applied, calculate (a) rotor current, (b) motor power output (including windage and friction), (c) rotor copper loss, and (d) slip at which maximum torque is developed.
Ans. (a) 91.57; (b) 23.89 kW; (c) 1.258 kW; (d) 0.14 (or 0.33 if only rotor circuit used)

5.40. A 3-phase, wye-connected, 12-pole induction motor is rated 500 hp, 220 V, 60 Hz. The stator resistance per phase is 0.4 Ω, the rotor resistance per phase in stator terms is 0.2 Ω, and the total rotor and stator reactance per phase in stator terms is 2 Ω. With rated voltage and frequency applied, the motor slip is 0.02. For this condition, find, on a per-phase basis, (a) the stator current (neglect magnetizing current), (b) the developed torque, (c) the rotor power input, (d) the rotor copper loss.
Ans. (a) 120 A; (b) 2292 N · m; (c) 144 kW; (d) 2880 W

5.41. A 3-phase, 12-pole induction motor is rated 500 hp, 2200 V, 60 Hz. At no-load, with rated voltage and frequency, the line current is 20 A and the input power is 14 kW. Assuming wye-connected windings, at 75°C the stator resistance per phase is 0.4 Ω, the rotor resistance per phase in stator terms is 0.2 Ω, and the equivalent reactance per phase $(X_1 + X_2')$ is 2 Ω. With rated voltage applied at rated frequency, the motor is loaded until its slip is 2%. For this condition (and for the given temperature), compute (a) the rotor current in stator terms, (b) the stator current, (c) the torque developed, (d) the power output, (e) the efficiency, and (f) the power factor. *Ans.* (a) 118.2 A; (b) 125 A; (c) 6644 N · m; (d) 410 kW (or 550 hp); (e) 91%; (f) 0.945 lagging

5.42. An approximate per phase equivalent circuit of a 3-phase, 220-V, wye-connected, 6-pole, 60-Hz induction motor is shown in Fig. 5-8(a) for which $R_1 = 0.3$ Ω; $X_1 = 0.4$ Ω; $R_2' = 0.4$ Ω; $X_2' = 0.6$ Ω; and $X_m = 15$ Ω. For the given numerical values calculate (a) the input current, power, and power factor per phase; and (b) the total developed (or electromagnetic) torque. The motor speed is 1080 rpm.
Ans. (a) 30.18 A; 10.302 kW; 0.896 lagging; (b) 75.45 N · m

5.43. An approximate per phase equivalent circuit of a three-phase induction motor is shown in Fig. 5-10. Test data on this motor are:

no-load test:	stator (applied) voltage = 120 V/phase
	input current = 5 A/phase
	input power = 480 W/phase

blocked-rotor test:	input voltage = 20 V/phase
	input current = 40 A/phase
	input power = 480 W/phase

Neglecting friction and windage losses, calculate the circuit parameters R_m, X_m, $(R_1 + R_2')$ and $(X_1 + X_2')$
Ans. 30 Ω; 40 Ω; 0.3 Ω; 0.4 Ω.

5.44. A certain 10-hp, 110-V, 3-phase, Y-connected, 60-Hz, 4-pole induction motor has the following test data:

no-load:	110 V, 21 A, 650 W
blocked-rotor:	24 V, 51 A, 1040 W
stator resistance *between terminals*:	0.08 Ω

(a) Determine the constants (per-phase values) of the approximate equivalent circuit (Fig. 5-10). (b) Determine the total friction, windage, and core loss of the motor.
Ans. (a) $R_m = 18.62$ Ω, $X_m = 5.45$ Ω, $R_1 = 0.04$ Ω, $R_2' = 0.36$ Ω, $X_1 + X_2' = 0.247$ Ω; (b) ≈632 W

5.45.   The motor of Problem 5.44 has a wound rotor. Assume that both stator and rotor windings are wye-connected. (*a*) If full voltage were applied to the stator with the rotor terminals short-circuited, how much starting current would the motor draw? (*b*) If the ratio of transformation per phase of the motor is 2, with the rotor having the greater number of turns, how much resistance should be added to each phase of the rotor to limit the starting current of the motor to 75 A when rated voltage is applied to the stator?
Ans. (*a*) 233.7 A;   (*b*) 2.72 Ω

5.46.   A 3-phase, 12-pole induction motor is rated 500 hp, 220 V, 60 Hz. At no-load, with rated voltage and frequency, the line current is 20 A and the power is 14 kW. Assuming wye-connected windings, the stator resistance per phase is 0.4 Ω, rotor resistance per phase in stator terms is 0.2 Ω, stator and rotor reactance per phase are each 1 Ω (in stator terms). With rated voltage and frequency applied, the motor is loaded until its slip is 2%. For this condition, calculate (*a*) the rotor current in stator terms, (*b*) the stator current, (*c*) the torque developed, (*d*) the power output, (*e*) the efficiency, and (*f*) the power factor. Use the approximate equivalent circuit (Fig. 5-10). (Notice that this is a repeat of Problem 5.41, but using a different circuit.)
Ans. (*a*) 120.0 A;   (*b*) 129.3 A;   (*c*) 6875 N · m;   (*d*) 409.4 kW (or 548 hp);   (*e*) 88.4%;
(*f*) 0.94 lagging

5.47.   For a given voltage supplied to the primary of a wye-delta starter (Problem 5.21) and the secondary changed from wye to delta, find the ratio of the (*a*) staring currents, (*b*) starting torques.
Ans.   (*a*) $I_Y/I_\Delta = 1/\sqrt{3}$ (phase values);   (*b*) $T_Y/T_\Delta = 1/3$

5.48.   A "Class B" induction motor has a lower starting current than a "Class A" because of its larger leakage inductance. How does this increase in leakage inductance affect (*a*) maximum motor torque, (*b*) slip at maximum torque, and (*c*) torque at rated load?   *Ans.* (*a*) lowers;   (*b*) lowers;   (*c*) no change

5.49.   Induction motors are often braked rapidly by a technique known as "plugging," which is the reversal of the phase sequence of the voltage supplying the motor. Assume that a motor with 4 poles is operating at 1750 rpm from an infinite bus (a load-independent voltage supply) at 60 Hz. Two of the stator supply leads are suddenly interchanged. (*a*) What is the new slip?   (*b*) What is the new rotor current frequency?   *Ans.* (*a*) 1.97;   (*b*) 118.33 Hz

5.50.   Using the circuit of Fig. 5-8(*a*), and neglecting the magnetizing reactance, show that for maximum developed power we must have

$$\frac{R_2'}{s}(1-s) = \sqrt{(R_1 + R_2')^2 + (X_1 + X_2')^2}$$

5.51.   (*a*) Show that the slip corresponding to the maximum power in Problem 5.50 is given by

$$s_m = \frac{R_2'}{R_2' + \sqrt{R_e^2 + X_e^2}}$$

where $R_e = R_1 + R_2'$ and $X_e = X_1 + X_2'$. (*b*) Determine the developed power at this slip, for a terminal

voltage V_1 (per phase).   *Ans.* (*b*) $\dfrac{V_1^2 \sqrt{R_e^2 + X_e^2}}{(R_e + \sqrt{R_e^2 + X_e^2})^2 + X_e^2}$ per phase

5.52. An induction motor is to be started at a reduced voltage such that the starting current will not exceed four times the full-load current, while developing a starting torque of 25% of the full-load torque. The full-load slip is 3%. Determine the factor by which the motor terminal voltage must be reduced at starting. *Ans.* 0.722

5.53. The double-cage rotor of an induction motor has the equivalent circuit shown in Fig. 5-17. The per-phase circuit parameters are $R_o = 10$, $R_i = 0.4$ Ω and $X_i = 4$, $X_o = 0.8$ Ω. Evaluate the ratio of the torques provided by the outer and the inner cages at starting. *Ans.* $T_o/T_i = 32$

5.54. The motor of Problem 5.53 runs with a slip of 4% on full-load at a rotor equivalent voltage of 50 V per phase. Calculate (a) the rotor current and (b) the torques contributed by the inner and outer cages. *Ans.* (a) 43 A; (b) $T_i = 1518$ N $\cdot$ m, $T_o = 250$ N $\cdot$ m

Chapter 6

Synchronous Machines

6.1 TYPES AND CONSTRUCTIONAL FEATURES

Synchronous machines are among the three most common types of electric machines; they are so called because they operate at constant speeds and constant frequencies under steady state. Like most rotating machines, a synchronous machine is capable of operating either as a motor or as a generator.

The operation of a synchronous generator is based on Faraday's law of electromagnetic induction, and a synchronous generator works very much like a dc generator, in which the generation of emf is by the relative motion of conductors and magnetic flux. However, a synchronous generator does not have a commutator as does a dc generator. The two basic parts of a synchronous machine are the *magnetic field structure*, carrying a dc-excited winding, and the *armature*. The armature often has a three-phase winding in which the ac emf is generated. Almost all modern synchronous machines have stationary armatures and rotating field structures. The dc winding on the rotating field structure is connected to an external source through slip rings and brushes. [Recall the construction of the elementary ac generator, Fig. 4-1(*a*), and the slip-ring-type induction motor, Chapter 5.] Some field structures do not have brushes, but instead have brushless excitation by rotating diodes. In some respects, the stator carrying the armature windings is similar to the stator of a polyphase induction motor (Fig. 5-3).

In addition to the armature and field windings, a synchronous machine has *damper bars* on the rotor. These come into play during transients and start-up (see Section 6.2).

Depending upon the rotor construction, a synchronous machine may be either a *round-rotor* type (Fig. 6-1) or a *salient-pole* type (Fig. 6-2). (Note that the armatures are not shown in Figs. 6-1 and 6-2.) The former type is used in high speed machines such as turbine generators, whereas the latter type is suitable for low-speed, waterwheel generators.

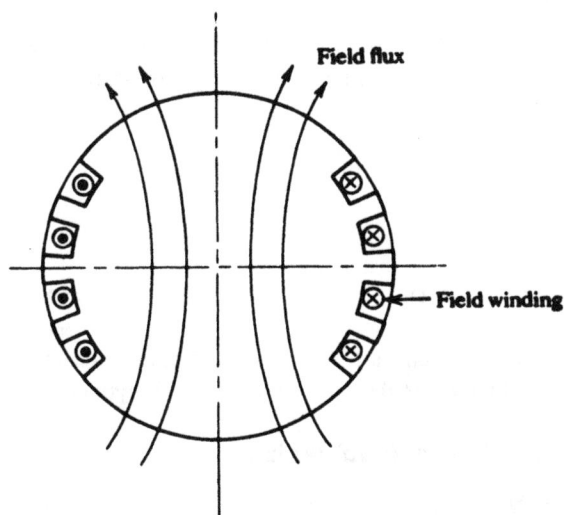

Fig. 6-1. Field winding on a round rotor.

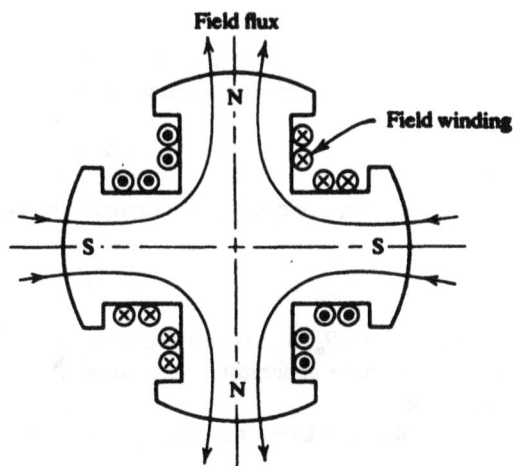

Fig. 6-2. Field winding on a salient rotor.

6.2 GENERATOR AND MOTOR OPERATION; THE EMF EQUATION

To understand the generator operation, we refer first to the 3-phase, round-rotor machine (Fig. 6-3), which has concentrated winding. It follows from (*4.2*) that the voltage induced in phase *A* is given by

$$v_A = V_m \sin \omega t$$

where ω is the angular velocity of the rotor. Phases B and C, being displaced from A and from each other by 120°, have voltages given by

$$v_B = V_m \sin (\omega t - 120°) \qquad v_C = V_m \sin (\omega t + 120°)$$

These voltages are sketched in Fig. 6-4. Hence, a 3-phase voltage is generated, of frequency $f = \omega/2\pi$ (Hz).

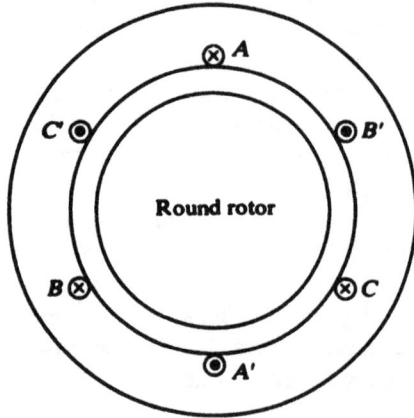

Fig. 6-3. A 3-phase, round-rotor,
synchronous machine.

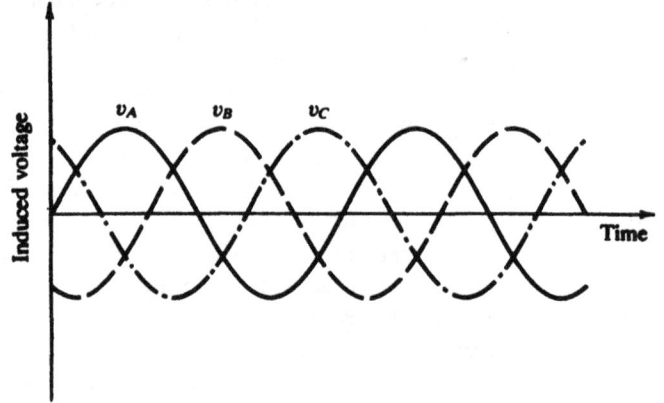

Fig. 6-4

Next, considering the salient-pole generator shown in Fig. 6-5, we let the flux-density distribution in the airgap, produced by the dc field winding, be

$$B(\theta) = B_m \cos \theta$$

where θ is measured with respect to the rotor axis, as shown in Fig. 6-5. Let the N-turn armature coil corresponding to phase A have radius r and axial length l. Then, when the rotor is in angular position $\alpha = \omega t$ (see Fig. 6-5), the flux linking the coil is

$$\lambda = N \int_{(\pi/2)-\alpha}^{(3\pi/2)-\alpha} B(\theta)\, lr\, d\theta = -2NB_m lr \cos \alpha$$

so that, by Faraday's law, the voltage induced in the coil is

$$v_A = \frac{d\lambda}{dt} = \frac{d\lambda}{d\alpha}\frac{d\alpha}{dt} = V_m \sin \omega t \qquad (6.1a)$$

where $V_m \equiv 2NB_m lr\omega$; similar expressions are found for phase B and phase C. We see that both round-rotor and salient-pole generators are governed by (6.1), which is known as the *emf equation* of a synchronous generator.

Reconsidering the amplitude of the induced voltage, V_m, form $(6.1a)$ we have

$$V_m = 2B_m lr\omega N \qquad (6.1b)$$

which is valid for a 2-pole machine. In general, if the machine has P poles (or $P/2$ pole pairs), $(6.1b)$ modifies to

$$V_m = \frac{2}{P}(2B_m lr\omega N) = \frac{4}{P} B_m lr\omega N \qquad (6.1c)$$

Clearly, when $P = 2$, $(6.1c)$ reduces to $(6.1b)$ as expected.

Turning now to the operation of a three-phase synchronous motor, and considering the salient-pole machine (Fig. 6-5), we observe that the three-phase armature (or stator) winding will produce a rotating magnetic field in the airgap, as in a three-phase induction motor (Section 5.3). The speed of rotation of the field, i.e., the synchronous speed n_s, is given by

$$n_s = \frac{120f}{p} \quad (\text{rpm}) \qquad\qquad (6.2)$$

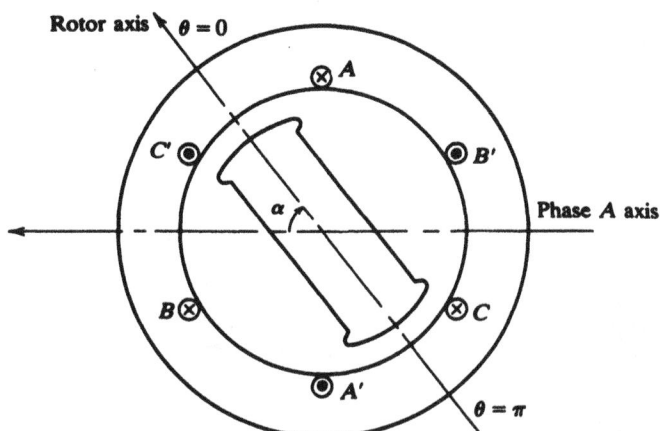

Fig. 6-5. A 3-phase, salient-pole, synchronous machine.

where p is the number of poles and f is the frequency of the voltage applied to the armature. However, with no short-circuited conductors on the rotor, the motor will not self-start. Suppose that the rotor of the salient-pole machine is brought to a speed close to n_s (by some auxiliary means). Then, even if there is no field excitation on the rotor, the rotor will align and rotate with the rotating field of the stator, because of the reluctance torque (Problem 3.24).

Clearly, no reluctance torque is present in the round-rotor machine of Fig. 6-3. Nevertheless, for either type of machine running at close to synchronous speed, if the field winding on the rotor is excited at such time as to place a north pole of the rotor field opposite a south pole of the stator field, then the two fields will lock in and the rotor will run at the synchronous speed.

In order to make the synchronous motor self-starting, it is provided with damper bars, which, like the cage of the induction motor, provide a starting torque. Once the rotor has pulled into step with the rotating stator field and runs at the synchronous speed, the damper bars go out of action. Any departure from the synchronous speed results in induced currents in the damper bars, which tend to restore the synchronous speed.

6.3 GENERATOR NO-LOAD, SHORT-CIRCUIT, AND VOLTAGE-REGULATION CHARACTERISTICS

The no-load or open-circuit voltage characteristic of a synchronous generator is similar to that of a dc generator. Figure 6-6 shows such a characteristic, with the effect of magnetic saturation included. Now, if the terminals of the generator are short-circuited, the induced voltage is dropped internally within the generator. The short-circuit current characteristic is also shown in Fig. 6-6. Expressed mathematically (on a per-phase basis):

$$\mathbf{V}_0 = \mathbf{I}_a \mathbf{Z}_s = \mathbf{I}_a (R_a + jX_s) \qquad\qquad (6.3)$$

In (6.3), $\mathbf{V}_0$ is the no-load armature voltage at a certain field current, and $\mathbf{I}_a$ is the short-circuit armature current at the same value of the field current. The impedance $\mathbf{Z}_s$ is known as the *synchronous impedance*; R_a is the armature resistance and X_s is defined as the *synchronous reactance*. The synchronous reactance is readily measured for a round-rotor generator, since it is independent of the rotor position in such a machine.

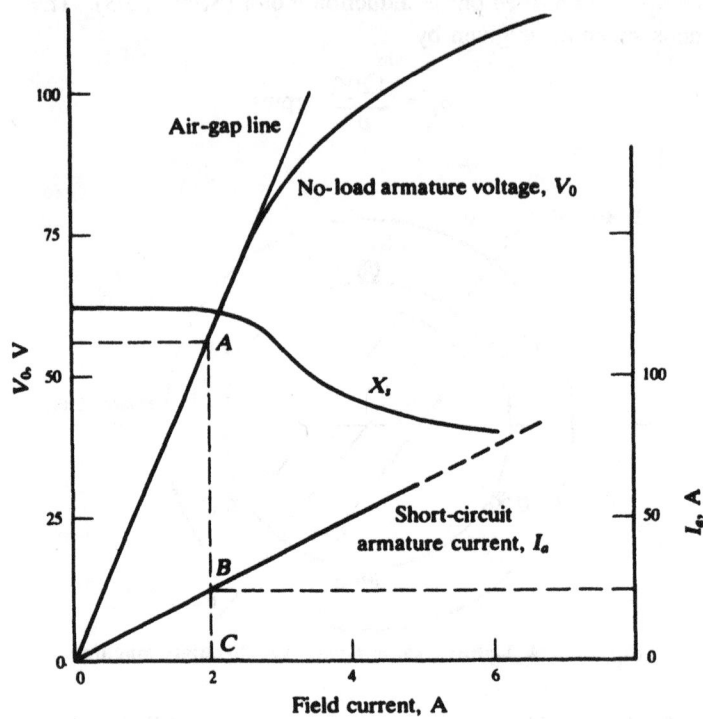

Fig. 6-6

In salient-pole generators, however, the synchronous reactance depends on the rotor position (see Section 6.6). In most synchronous machines, $R_a \ll X_s$, so that, in terms of Fig. 6-6,

$$X_s \approx Z_s = \frac{\overline{AC}}{\overline{BC}}$$

Thus X_s varies with field current as indicated by the falling (because of saturation) curve in Fig. 6-6. However, for most calculations, we shall use the linear (constant) value of X_s.

As for a transformer or a dc generator, we define the *voltage regulation* of a synchronous generator at a give load as

$$\text{percent voltage regulation} \equiv \frac{V_0 - V_t}{V_t} \times 100\% \qquad (6.4)$$

where V_t is the terminal voltage per phase on load and V_0 is the no-load terminal voltage per phase. Knowing X_s (for a round-rotor generator) and V_t, we can find V_0 from (6.3) and hence determine the voltage regulation.

Unlike what happens in a dc generator, the voltage regulation of a synchronous generator may become zero or even negative, depending upon the power factor and the load (see Problems 6.9, 6.10, and 6.11). Neglecting the armature resistance, we show phasor diagrams for lagging and leading power factors in Fig. 6-7.

6.4 POWER-ANGLE CHARACTERISTIC OF A ROUND-ROTOR MACHINE

With reference to Fig. 6-7, ϕ is the power-factor angle, and δ, the angle by which V_0 leads V_t, is defined as the *power angle*. To justify this name, we obtain from Fig. 6-7:

$$I_a X_s \cos \phi = V_0 \sin \delta \qquad (6.5)$$

where it is assumed that $\delta > 0$ (generator action). But the power developed (per phase) by the generator, P_d, is the power supplied to the load. Thus,

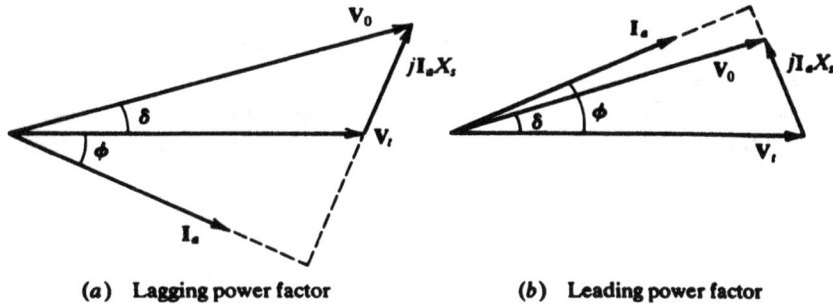

(a) Lagging power factor   (b) Leading power factor

Fig. 6-7. Phasor diagrams for a synchronous generator.

$$\text{generator:} \quad P_d = V_t I_a \cos \phi \tag{6.6}$$

Comparing (6.5) and (6.6) yields

$$P_d = \frac{V_0 V_t}{X_s} \sin \delta \tag{6.7}$$

which shows that the power developed by the generator is proportional to $\sin \delta$.

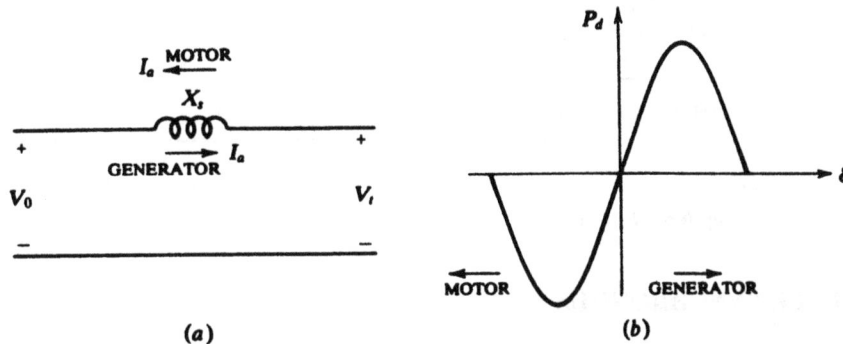

(a)   (b)

Fig. 6-8. For round-rotor machines.

As is indicated in Fig. 6-8(a), a (round-rotor) *motor consumes* electrical power in the amount $V_t I_a \cos \phi$ per phase, if armature resistance is neglected. We therefore define the power *developed by* the motor as

$$\text{motor:} \quad P_d = -V_t I_a \cos \phi$$

With this understanding, (6.7) is also valid for a round-rotor motor, where now δ, and hence $\sin \delta$, is negative (V_t leads V_0). In short, (6.7) is the power-angle characteristic of a round-rotor synchronous machine; a graph is given in Fig. 6-8(b).

6.5 PERFORMANCE OF THE ROUND-ROTOR MOTOR

For a motor, Fig. 6-8(a) gives

$$\mathbf{V}_t = \mathbf{V}_0 + j\mathbf{I}_a X_s \tag{6.8}$$

If the motor operates at constant power, then (6.5) and (6.7) imply that

$$V_0 \sin \delta = I_a X_s \cos \phi = \text{constant} \qquad (6.9)$$

for a given terminal voltage V_t.

Now, V_0 depends upon the field current, I_f. Consider two cases: (1) I_f is so adjusted that $V_0 < V_t$ (the machine is *underexcited*) and (2) I_f is increased to a point where $V_0 > V_t$ (the machine is *overexcited*). The voltage-current relationships for the two cases are shown in Fig. 6-9(*a*), in which single primes refer to underexcited, and double primes to overexcited, operation. At constant power, δ is less negative for $V_0 > V_t$ than for $V_0 < V_t$, as governed by (6.9). Notice that an underexcited motor operates at a lagging power factor (I_a lagging V_t), whereas an overexcited motor operates at a leading power factor. Thus, *the operating power factor of the motor is controlled by varying the field excitation* (thereby altering V_0); this is a very important property of synchronous motors. The locus of the armature current at a constant load, as given by (6.9), for varying field current is shown in Fig. 6-9(*a*). From this we can obtain the variation of the armature current, I_a, with the field current, I_f (corresponding to V_0); the results for several different loads are plotted in Fig. 6-9(*b*). These curves are known as the *V-curves* of the synchronous motor. One of the applications of a synchronous motor is in power-factor correction.

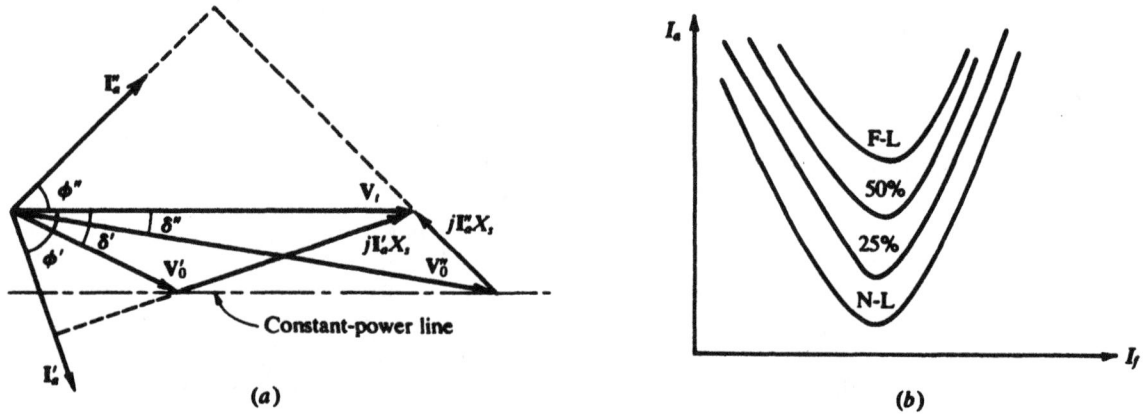

Fig. 6-9. **Round-rotor motor operation at constant power.**

6.6 SALIENT-POLE SYNCHRONOUS MACHINES

We have alluded to the unexcited salient-pole machine, namely, the reluctance motor Problem 3.24. In Fig. 3-17(*b*), we implicitly defined the *direct-* and *quadrature-axis inductances*, L_d and L_q, as the values of the inductance when the rotor and stator axes are aligned and when they are antialigned. Corresponding, we may define the *d-axis* and *q-axis synchronous reactances*, X_d and X_q, for a salient-pole synchronous machine. Thus, for generator operations, we draw the phasor diagram of Fig. 6-10. Notice that $\mathbf{I}_a$ has been resolved into its *d-* and *q-axis* (fictitious) components, $\mathbf{I}_d$ and $\mathbf{I}_q$. With the help of this phasor diagram, we obtain:

$$I_d = I_a \sin(\delta + \phi) \qquad I_q = I_a \cos(\delta + \phi) \qquad (6.10)$$

$$V_t \sin \delta = I_q X_q = I_a X_q \cos(\delta + \phi) \qquad (6.11)$$

Expansion of (*6.11*) gives

$$\tan \delta = \frac{I_a X_q \cos \phi}{V_t + I_a X_q \sin \phi} \qquad (6.12)$$

With δ known (in terms of ϕ), the voltage regulation may be computed from

$$V_0 = V_t \cos \delta + I_d X_d$$

$$\text{percent regulation} = \frac{V_0 - V_t}{V_t} \times 100\%$$

In fact, the phasor diagram depicts the complete performance characteristics of the machine.

Example 6.1 Let us use Fig. 6-10 to derive the power-angle characteristics of a salient-pole generator. If armature resistance is neglected, (6.6) applies. Now, from Fig. 6-10, the projection of I_a on V_t is

$$\frac{P_d}{V_t} = I_a \cos \phi = I_q \cos \delta + I_d \sin \delta \qquad (6.13)$$

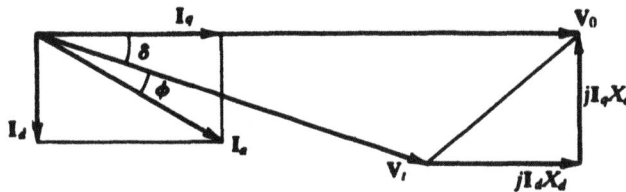

Fig. 6-10. Phasor diagram of a salient-pole generator.

Solving

$$I_q X_q = V_t \sin \delta \quad \text{and} \quad I_d X_d = V_0 - V_t \cos \delta$$

for I_q and I_d, and substituting in (6.13), gives

$$P_d = \frac{V_0 V_t}{X_d} \sin \delta + \frac{V_t^2}{2} \left(\frac{1}{X_q} - \frac{1}{X_d} \right) \sin 2\delta \qquad (6.14)$$

Equation (6.14) can also be established for a salient-pole motor ($\delta < 0$); the graph of (6.14) is given in Fig. 6-11. Observe that for $X_d = X_q = X_s$, (6.14) reduces to the round-rotor equation, (6.7).

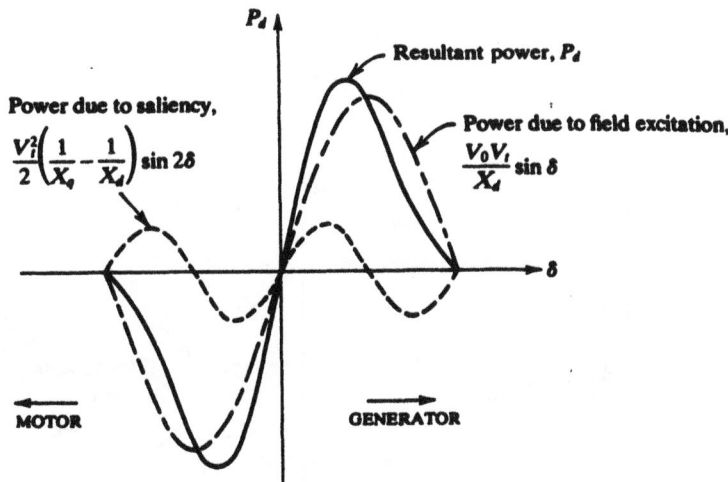

Fig. 6-11. For salient-pole machines.

6.7 TRANSIENTS IN SYNCHRONOUS MACHINES

Sudden Short-Circuit at the Armature Terminals

Consider a three-phase generator, on no-load, running at its synchronous speed and carrying a constant field current. Suddenly, the three phases are short-circuited. Symmetrical short-circuit armature current is graphed in Fig. 6-12. Notice that for the first few cycles the current, i_a, decays very rapidly; we term this duration the *subtransient period*. During the next several cycles, the current decreases somewhat slowly, and this range is called the *transient period*. Finally, the current reaches its steady-state value. These currents are respectively limited by the *subtransient reactance*, x_d''; the *transient reactance*, x_d'; and the synchronous reactance, X_d (or X_s). The subtransient reactance is essentially due to the presence of damper bars; the transient reactance accounts for the field winding; and the synchronous reactance is reactance due to the armature windings. It can be shown that the envelope of the instantaneous armature current (dashed curves in Fig. 6-12) is given by

$$i_a^{\cdot} = \pm V_0 \left[\left(\frac{1}{x_d''} - \frac{1}{x_d'} \right) e^{-t/\tau_d} + \left(\frac{1}{x_d'} - \frac{1}{X_d} \right) e^{-t/\tau_d'} + \frac{1}{X_d} \right] \tag{6.15}$$

where
$\tau_d'' \equiv$ subtransient time constant
$\tau_d' \equiv$ transient time constant
$V_0 \equiv$ open-circuit armature phase voltage

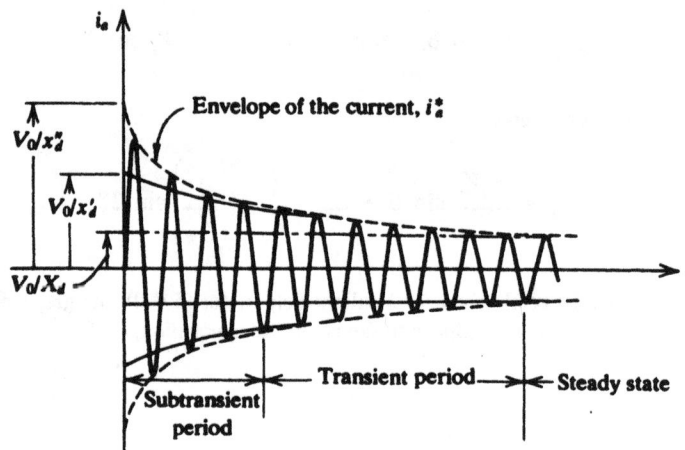

Fig. 6-12

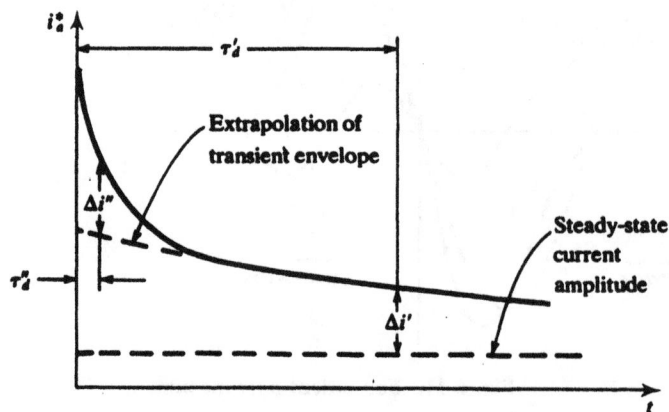

Fig. 6-13

The upper branch of the envelope is separately shown in Fig. 6-13.

The reactances and the time constants can be determined from design data, but the details are extremely cumbersome. On the other hand, these may be determined from test data and Figs. 6-12 and 6-13, as illustrated by Problem 6.30. In Fig. 6-13,

$$\Delta i'' = 0.368\ (i_d'' - i_d') \qquad \Delta i' = 0.368\ (i_d' - i_d)$$

where $i_d'' \equiv V_0/x_d''$, $i_d' \equiv V_0/x_d'$, and $i_d \equiv V_0/X_d$. Table 6-1 gives typical values of synchronous machine constants; the per-unit values are based on the machine rating.

Table 6-1. Per-unit synchronous machine reactances and time constants

Constant	Salient-Pole Machine	Round-Rotor Machine
X_d (Ω)	1.0 to 1.25	1.0 to 1.2
X_q	0.65 to 0.80	1.0 to 1.2
x_d'	0.35 to 0.40	0.15 to 0.25
x_d''	0.20 to 0.30	0.10 to 0.15
τ_d (s)	0.15	0.15
τ_d'	0.9 to 1.1	1.4 to 2.0
τ_d''	0.03 to 0.04	0.03 to 0.04

Sudden Change in Load

The mechanical equation of motion of a synchronous generator is

$$J\ \frac{d^2\theta_m}{dt^2} + b\ \frac{d\theta_m}{dt} + T_e = T_m \tag{6.16}$$

where T_e is the electromagnetic torque developed by the machine; T_m is an opposing externally applied torque; J is the moment of inertia of the rotating parts (the rotor and the prime mover); and b is the friction coefficient, including electrical damping.

Let us consider a 3-phase, 2-pole, round-rotor machine, and assume that the frequency of mechanical oscillations (because of some disturbance) is small, so that the steady-state power-angle characteristic can be used. The total power developed by the machine is given by (6.7) as

$$\frac{3V_0V_t}{X_s}\ \sin\delta = T_e\omega_m \tag{6.17}$$

where V_0, V_t, and X_s are per-phase values; and where ω_m, the angular velocity of the rotor, is the same as the synchronous angular velocity under steady-state conditions. Let a sudden load change in the form of a small impulsive torque, ΔT_m, produce changes $\Delta\theta_m$ and ΔT_e in θ_m and T_e, respectively. Then (6.16) yields

$$J\frac{d^2(\Delta\theta_m)}{dt^2} + b\ \frac{d(\Delta\theta_m)}{dt} + \Delta T_e = \Delta T_m \tag{6.18}$$

From (6.17), the change in electromagnetic torque is

$$\Delta T_e = \frac{3V_0V_t}{\omega_m X_s}\ \Delta\ (\sin\delta) \approx \frac{3V_0V_t}{\omega_m X_s}\ \frac{d(\sin\delta)}{d\delta} = \frac{3V_0V_t\ \cos\delta}{\omega_m X_s}\ \Delta\delta \tag{6.19}$$

[A result equivalent to (6.19) is found in Problem 6.25(b), where $\psi = \Delta\delta$.] We denote the multiplier of $\Delta\delta$ on the right of (6.19) as the *torque constant*, k_e, the power angle is given its steady-state value. Finally, we observe that $\Delta\theta_m = \Delta\delta$ for a two-pole machine (see Problem 6.26). Substituting for $\Delta\theta_m$ and ΔT_e in (6.18),

$$J\ \frac{d^2(\Delta\delta)}{dt^2} + b\ \frac{d(\Delta\delta)}{dt} + k_e\ \Delta\delta = \Delta T_m \tag{6.20}$$

which is a linear second-order differential equation for $\Delta\delta$. If we compare (6.20) with the second order differential equation of a mechanical system, the natural frequency of oscillation and the damping ratio are found to be

$$f_n = \frac{1}{2\pi}\sqrt{\frac{k_e}{J}} \quad \text{(Hz)} \qquad \zeta = \frac{b}{2\sqrt{k_e J}} \qquad\qquad (6.21)$$

In most machines, $0.2 \text{ Hz} \le f_n \le 2.0 \text{ Hz}$.

Solved Problems

6.1.   A 4-pole induction motor, running with 5% slip, is supplied by a 60-Hz synchronous generator. (a) Calculate the speed of the motor. (b) What is the generator speed if it has six poles?

(a) $n = (1 - s)n_s = (1 - 0.05)\dfrac{120(60)}{4} = 1710 \text{ rpm}$

(b) $n_s = \dfrac{120f}{p} = \dfrac{120(60)}{6} = 1200 \text{ rpm}$

6.2.   For a 60-Hz synchronous generator, list six possible combinations of number of poles and speed.

From

$$f = \frac{pn_s}{120}$$

we must have $pn_s = 120(60) = 7200$ rpm. Hence Table 6-2.

Table 6-2

No. of Poles	Speed, rpm
2	3600
4	1800
6	1200
8	900
10	720
12	600

6.3.   The flux-density distribution in the airgap of a 60-Hz, 4-pole, salient-pole machine is sinusoidal, having an amplitude of 0.6 T. Calculate the instantaneous and rms values of the voltage induced in a 150-turn coil on the armature, if the axial length of the armature and its inner diameter are both 100 mm.

Substituting the numerical values, in (*6.1c*) we obtain

$$v = \frac{4}{4}(150)(0.6)(0.100)(0.50)(377) \sin 377t = 169.65 \sin 377t \quad \text{(V)}$$

and
$$V = \frac{169.65}{\sqrt{2}} = 119.96 \text{ V}$$

6.4. The flux-density distribution produced in a synchronous generator by an ac-excited field winding is

$$B(\theta, t) = B_m \sin \omega_1 t \cos \theta$$

Find the armature voltage induced in an *N*-turn coil if the rotor (or field) rotates at ω_2 (rad/s). Comment on the special case $\omega_1 = \omega_2 = \omega$.

Proceeding as in the derivation of (*6.1*),

$$\lambda = -2NlrB_m \sin \omega_1 t \cos \omega_2 t$$

Thus,

$$v = \frac{d\lambda}{dt} = -2NlrB_m (\omega_1 \cos \omega_1 t \cos \omega_2 t - \omega_2 \sin \omega_1 t \sin \omega_2 t)$$

When $\omega_1 = \omega_2 = \omega$, this reduces to

$$v = -2NlrB_m \omega \cos 2\omega t$$

i.e., a double-frequency generator.

6.5. A 4-pole, 3-phase synchronous machine has 48 slots. Calculate the armature-winding distribution factor.

Recall from Problem 5.1 that

$$k_d = \frac{\sin (q\alpha/2)}{q \sin (\alpha/2)}$$

In this case,

$$q \equiv \text{slots/pole/phase} = \frac{48}{(4)(3)} = 4 \quad \text{and} \quad \alpha = \frac{180°}{mq} = \frac{180°}{(3)(4)} = 15°$$

where *m* ≡ number of phases = 3. Hence

$$k_d = \frac{\sin 30°}{4 \sin 7.5°} = 0.958$$

The distribution factors for a few three-phase windings follow.

q	2	3	4	5	6	8	∞
k_d	0.966	0.960	0.958	0.957	0.957	0.956	0.955

6.6. A 3-phase, 8-pole, 60-Hz, wye-connected, salient-pole synchronous generator has 96 slots, with 4 conductors per slot connected in series in each phase. The coil pitch is 10 slots. If the maximum value of the airgap flux is 60 mWb and the flux-density distribution in the airgap is sinusoidal, determine (*a*) the rms phase voltage and (*b*) the rms line voltage. (*c*) If each phase is capable of carrying 650 A in current, what is the kVA rating of the machine?

(*a*)   The rms phase voltage is given by

$$E_{rms} = k_w \ (4.44 f N \phi_m) \quad (V)$$

[from (6.1), with $\dfrac{2}{P} B_m (2lr) = \dfrac{2}{P} B_m A = \phi_m$ and $\omega = 2\pi f$], where $k_w = k_p k_d$ is the winding factor (Problem 5.3). As in Problem 6.5, we have $q = 4$, so that $k_d = 0.958$. The pitch factor was found in Problem 5.2 to be

$$k_p = \sin \frac{\pi \beta}{2\tau}$$

In this case,

$$\tau \equiv \text{pitch of full coil} = \frac{96}{8} = 12 \text{ slots}$$

$$\beta \equiv \text{pitch of fractional coil} = 10 \text{ slots}$$

whence

$$k_p = \sin \frac{10\pi}{24} = 0.966$$

The number of turns per phase, *N*, is half the number of conductors per phase:

$$N = \frac{1}{2}\left(\frac{96}{3}\right)(4) = 64$$

Thus,

$$E_{rms} = \frac{2}{8} \ (0.966)(0.958)(4.44)(60)(64)(60 \times 10^{-3}) = 236.7 \text{ V}$$

(*b*)   rms line voltage = $\sqrt{3}$ (236.7) = 410 V

(*c*)   machine rating = 3 × 0.2367 × 650 = 461.565 kVA

6.7.   If the machine of Problem 6.6 has 6% third harmonic present in the airgap flux density and has a full-pitch winding, what is the rms value of the phase voltage?

The rms value of the fundamental voltage is

$$E_{1 \ rms} = k_2(4.44 \ f N \phi_m) = (0.958)(4.44)(60)(64)(60 \times 10^{-3}) = 980 \text{ V}$$

For the third harmonic, the slot angle is effectively trebled, i.e.,

$$\alpha = \frac{540°}{mq} = \frac{540°}{(3)(4)} = 45°$$

so that

$$k_{d3} = \frac{\sin 90°}{4 \sin 22.5°} = 0.65$$

Now, if we rederive (6.1), this time starting with $B_3(\theta) = B_{m3} \cos 3\theta$, we find that the expression for λ includes an extra factor 1/3. Therefore, in effect,

$$\phi_{m3} = \frac{2}{8} \times \frac{1}{3} (0.06)(60 \times 10^{-3}) = 0.3 \times 10^{-3} \text{ Wb}$$

and

$$E_{3\,rms} = k_{d3} (4.44 \, f_3 N \phi_{m3}) = (0.65)(4.44)(3 \times 60)(64)(0.3 \times 10^{-3}) = 10 \text{ V}$$

and

$$E_{rms} = \sqrt{E_{1\,rms}^2 + E_{3\,rms}^2} = \sqrt{(236.7)^2 + (10)^2} = 236.9 \text{ V}$$

6.8.   A 3-phase, 60-Hz, 2-pole wye-connected, armature winding of a generator has 6 slots per pole per phase. The pole pitch is 10 slots and the coil pitch is 9 slots. The winding is double-layer and has 30 turns per phase. If the airgap flux is sinusoidally distributed, what must be its maximum value to give 600 V across the lines?

For $q = 6$, $k_d = 0.957$ (Problem 6.5); and for $\beta = 9$, $\tau = 10$,

$$k_p = \sin \frac{9\pi}{20} = 0.987$$

Then, as in Problem 6.6(a),

$$\frac{600}{\sqrt{3}} = (0.987)(0.957)(4.44)(60)(30)\phi_m$$

whence, $\phi_m = 45.89$ mWb.

6.9.   A 3-phase, wye-connected, round-rotor synchronous generator rated at 10 kVA, 230 V has a synchronous reactance of 1.2 Ω per phase and an armature resistance of 0.5 Ω per phase. Calculate the percent voltage regulation at full-load with 0.8 lagging power factor.

The phasor diagram is shown in Fig. 6-14, from which (ϕ is negative):

$$V_0 = \sqrt{(V_t \cos \phi + I_a R_a)^2 + (|V_t \sin \phi| + I_a X_s)^2}$$

Substituting

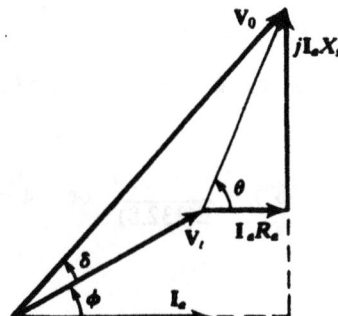

Fig. 6-14

$$V_t = \frac{230}{\sqrt{3}} = 132.8 \text{ V} \qquad I_a = \frac{(10 \times 10^3)/3}{230/\sqrt{3}} = 25.1 \text{ A}$$

and the other data yields

$$V_0 = \sqrt{(106.24 + 12.55)^2 + (79.68 + 30.12)^2} = 161.76 \text{ V}$$

Then

$$\frac{V_0 - V_t}{V_t} \times 100\% = \frac{161.76 - 132.8}{132.8} \times 100\% = 21.8\%$$

6.10. Repeat Problem 6.9 for the case of 0.8 leading power factor, other data remaining unchanged.

Let $\mathbf{V}_t = 132.8 + j0$ V be the reference phasor. Then,

$$\mathbf{I}_a = 25.1(0.8 + j0.6) \text{ A}$$

$$\mathbf{Z}_s = 0.5 + j1.2 \ \Omega$$

$$\mathbf{I}_a\mathbf{Z}_s = -8 + j31.6 \text{ V}$$

$$\mathbf{V}_0 = \mathbf{V}_t + \mathbf{I}_a\mathbf{Z}_s = 124.8 + j31.6 \text{ V}$$

or $V_0 = 128.7$ V. Hence,

$$\text{percent regulation} = \frac{128.7 - 132.8}{132.8} \times 100\% = -3.1\%$$

Notice that this problem has been solved without using the phasor diagram, and that the voltage regulation at full-load with 0.8 leading power factor is negative.

6.11. For the generator of Problem 6.9, determine the power factor such that the voltage regulation is zero on full-load.

Let ϕ be the required power-factor angle, so that $\mathbf{I}_a = 25.1\angle\phi$ A. Then,

$$\mathbf{Z}_s = 0.5 + j1.2 = 1.3\angle67.38° \ \Omega$$

$$\mathbf{I}_a\mathbf{Z}_s = 32.63\angle(\phi + 67.38°) = 32.63 \cos (\phi + 67.38°) + j32.63 \sin (\phi + 67.38°) \text{ V}$$

$$\mathbf{V}_0 = 132.8 + j0 + 32.63 \cos (\phi + 67.38°) + j32.63 \sin (\phi + 67.38°) \text{ V}$$

For zero voltage regulation, $V_0 = V_t = 132.8$ V, i.e.,

$$(132.8)^2 = [132.8 + 32.63 \cos (\phi + 67.38°)]^2 + [32.63 \sin (\phi + 67.38°)]^2$$

which gives

$$\cos (\phi + 67.38°) = \frac{-32.63}{2(132.8)} \quad \text{or} \quad \phi = +29.67°$$

Thus, $\cos \phi = 0.869$ leading.

6.12. A certain 3-phase, round-rotor synchronous generator is rated 150 MW, 0.85 power factor, 12.6 kV, 60 Hz, and 1800 rpm. Each winding has a line-to-neutral resistance of 1.535 mΩ. The data for the no-load magnetization curve are:

Field current, A	200	300	400	500	600	700	800	900
Armature voltage (line-to-line), kV	3.8	5.8	7.8	9.8	11.3	12.6	13.5	14.2

The short-ciruit armature-current test gives a straight line through the origin and through rated armature current at 700-A field current. (a) Determine the unsaturated synchronous impedance per phase. (b) Determine the saturated synchronous impedance per phase. (c) Draw a phasor diagram and determine the voltage regulation for the condition of rated load and 0.85 pf lagging. (d) Repeat part (c) for rated load and 0.85 pf leading. The saturation curve and the short-circuit test data are plotted in Fig. 6-15.

$$(I_a)_{rated} = \frac{(150 \times 10^6)/3}{[(12.6 \times 10^3)/\sqrt{3}](0.85)} = 8086 \text{ A}$$

From Fig. 6-15:

(a) $$Z_s \text{ (unsaturated)} = \frac{7.8/\sqrt{3}}{4.6} = 0.979 \ \Omega/\text{per phase} \approx X_s$$

(b) $$Z_s \text{ (saturated)} = \frac{13.5/\sqrt{3}}{9.2} = 0.847 \ \Omega \text{ per phase} \approx X_s$$

(c) $\phi = \cos^{-1} 0.85 = -31.8°$. The phasor diagram is shown in Fig. 6-16(a). Hence,

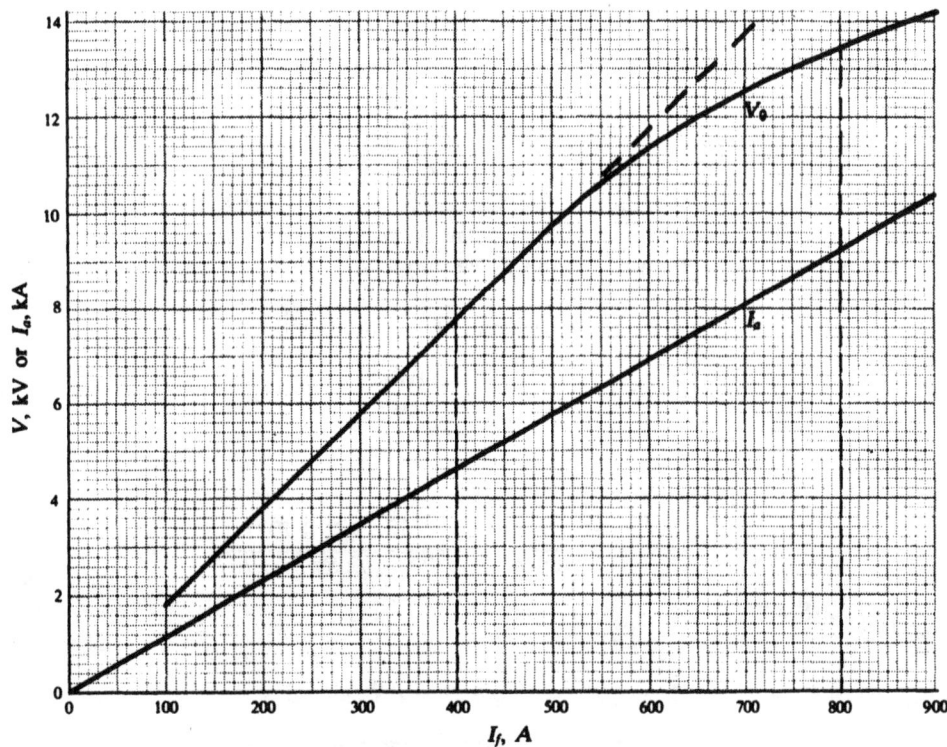

Fig. 6-15

$$\mathbf{V}_0 = \frac{12.6 \times 10^3}{\sqrt{3}} + (8086)(1.535 \times 10^{-3}) \angle -31.8° + (8086)(0.979) \angle 58.2°$$

$$= 7275 + (10.6 - j6.5) + (4171 + j6728) = 11\,455 + j6721 = 13\,281 \angle 30.4° \text{ V}$$

(The fact that use of 0.979, the unsaturated synchronous reactance, leads to a value of V_0 that is definitely in the saturated region of the magnetization curve indicates that an iterative procedure should have been used to find $\mathbf{V}_0$. However, we shall let the result stand as a first approximation.)

$$\text{regulation} = \frac{13\,281 - 7275}{7275} = 82.6\%$$

(*d*)   For $\phi = +31.8°$, Fig. 6-16(*b*) gives the phasor diagram. Thus,

$$\mathbf{V}_0 = 7275 + (10.6 + j6.5) + (-4171 + j6728) = 3115.6 + j6734.5 = 7420 \angle 65.2° \text{ V}$$

$$\text{regulation} = \frac{7420 - 7275}{7275} = 2.0\%$$

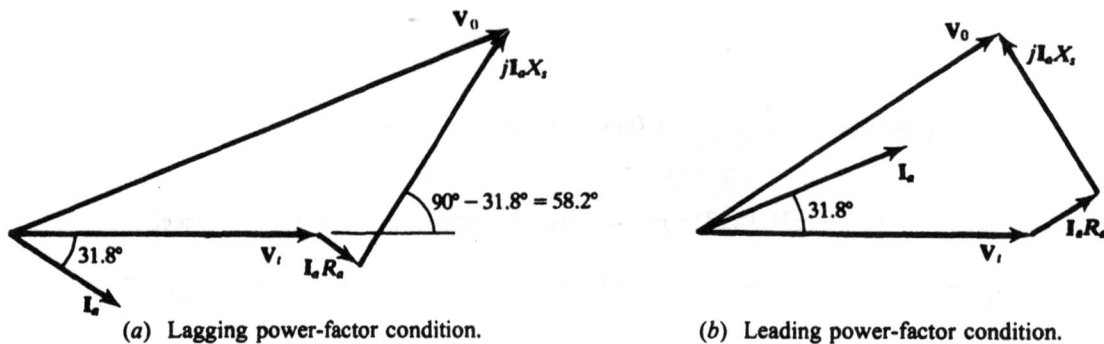

(*a*) Lagging power-factor condition.   (*b*) Leading power-factor condition.

Fig. 6-16

6.13.   A 20-kVA, 220-V, wye-connected, 3-phase, salient-pole synchronous generator supplies rated load at 0.707 lagging power factor. The reactances per phase are $X_d = 2X_q = 4 \ \Omega$. Neglecting the armature resistance, determine (*a*) the power angle and (*b*) the percent voltage regulation.

(*a*)   From (*6.12*) and the phasor diagram of Fig. 6-10, with

$$V_t = \frac{220}{\sqrt{3}} = 127 \text{ V} \qquad I_a = \frac{(20 \times 10^3)/3}{220/\sqrt{3}} = 52.5 \text{ A} \qquad \phi = \cos^{-1} 0.707 = 45°$$

we get

$$\tan \delta = \frac{I_a X_q \cos \phi}{V_t + I_a X_q \sin \phi} = \frac{(52.5)(2)(0.707)}{127 + (52.5)(2)(0.707)} = 0.369$$

or $\delta = 20.25°$.

(*b*) $V_0 = V_t \cos \delta + I_d X_d = V_t \cos \delta + I_a X_d \sin (\delta + \phi)$

$$= 127 \cos 20.25° + (52.5)(4)\sin (20.25° + 45°) = 309.8 \text{ V}$$

$$\text{percent voltage regulation} = \frac{309.8 - 127}{127} \times 100\% = 144\%$$

6.14. (a) Determine the power developed by the generator of Problem 6.13 and verify that it is equal to the power supplied to the load. (b) How much power is developed due to saliency?

(a)   The power developed per phase is given by (6.14) as

$$P_d = \frac{(309.8)(127)}{4} \sin 20.25° + \frac{(127)^2}{2} \left(\frac{1}{2} - \frac{1}{4} \right) \sin 40.50°$$

$$= 3404.46 + 1309.37 = 4713.8 \text{ W}$$

The power supplied to the load per phase is $(20 \times 10^3)(0.707)/3 = 4713.3$ W.

(b)   From (a), the power due to saliency is $3(1309.37) = 3928$ W

6.15. A 3-phase, salient-pole, 440-V, wye-connected synchronous generator operates at a power angle of 20° while developing a power of 36 kW. The machine constants per phase are $X_d = 2.5$, $X_q = 5$ Ω; R_a is negligible. Calculate the voltage regulation for the given operating conditions.

$$V_t = \frac{440}{\sqrt{3}} = 254 \text{ V} \qquad \delta = 20° \qquad P_d = \frac{36 \times 10^3}{3} = 12\,000 \text{ W}$$

From (6.14) we get

$$12\,000 = \frac{V_0(254)}{5} \sin 20° + \frac{(254)^2}{2} \left(\frac{1}{2.5} - \frac{1}{5} \right) \sin 40° \quad \text{or} \quad V_0 = 452 \text{ V}$$

$$\text{regulation} = \frac{452 - 254}{254} = 78\%$$

6.16. A 3-phase, wye-connected load takes 50 A in current at 0.707 lagging power factor with 220 V between the lines. A 3-phase, wye-connected, round-rotor synchronous motor, having a synchronous reactance of 1.27 Ω per phase, is connected in parallel with the load. The power developed by the motor is 33 kW at a power angle of 30°. Neglecting the armature resistance, calculate (a) the reactive power (in kvar) of the motor and (b) the overall power factor of the motor and the load.

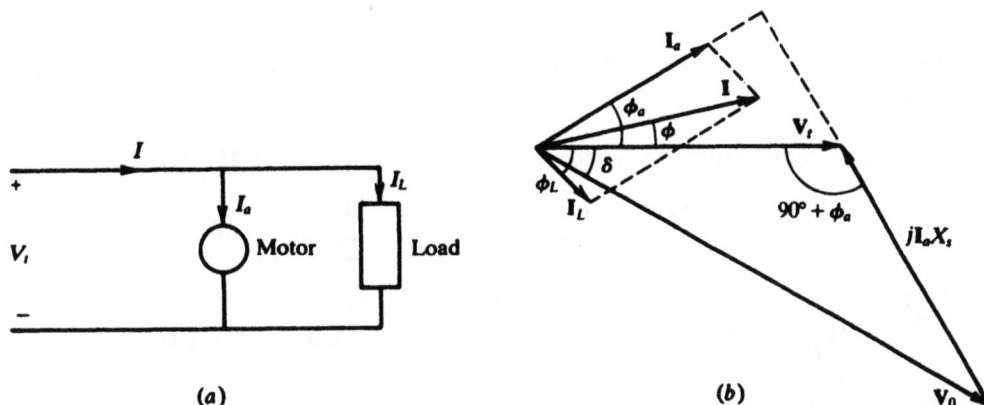

Fig. 6-17

(a)   The circuit and the phasor diagram, on a per-phase basis, are shown in Fig. 6-17. From (6.7),

$$P_d = \frac{33 \times 10^3}{3} = \frac{220}{\sqrt{3}} \frac{V_0}{1.27} \sin 30° \quad \text{or} \quad V_0 = 220 \text{ V}$$

By the parallel connection, Fig. 6-17(a), $I_a X_s = V_t = 220/\sqrt{3}$ V. Then, from the isosceles triangle in Fig. 6-17(b),

$$2\delta + 90° - \phi_a = 180° \quad \text{or} \quad \phi_a = 90 - 2\delta = 30°$$

$$\text{motor reactive power} = 3V_t I_a \sin \phi_a = 3 \frac{V_t^2}{X_s} \sin \phi_a$$

and

$$= 3 \frac{(220/\sqrt{3})^2}{1.27} \left(\frac{1}{2}\right) = 19000 \text{ var} = 19 \text{ kvar}$$

(b)   From Fig. 6-17(b), the projection of **I** on **V**$_t$ is

$$I \cos \phi = I_a \cos \phi_a + I_L \cos \phi_L$$

and its projection perpendicular to **V**$_t$ is (taking account of the fact that ϕ_L is negative)

$$I \sin \phi = I_a \sin \phi_a + I_L \sin \phi_L$$

Hence

$$\tan \phi = \frac{I_a \sin \phi_a + I_L \sin \phi_L}{I_a \cos \phi_a + I_L \cos \phi_L}$$

Substituting the values

$$I_a = \frac{220/\sqrt{3}}{1.27} = 100 \text{ A} \quad I_L = 50 \text{ A} \quad \phi_a = 30° \quad \phi_L = -45°$$

we obtain

$$\tan \phi = 0.120 \quad \text{or} \quad \cos \phi = 0.993 \text{ leading}$$

6.17. Including the effect of armature resistance, R_a, show that the power developed by a round-rotor synchronous motor is given (on a per-phase basis) by

$$P_d = -\frac{V_0 V_t}{Z_s} \cos(\delta + \theta) + V_0^2 \frac{R_a}{Z_s^2}$$

where V_t is the motor terminal voltage, V_0 is the internal (or induced) voltage, Z_s is the synchronous impedance, θ is the impedance angle, and δ is the power angle ($\delta < 0$).

For motor operation, the appropriate phasor diagram is Fig. 6-20. We have:

$$\mathbf{I}_a = \frac{\mathbf{V}_t - \mathbf{V}_0}{\mathbf{Z}_s} = \frac{V_t \angle 0° - V_0 \angle \delta}{Z_s \angle \theta} = \frac{V_t}{Z_s} \angle -\theta - \frac{V_0}{Z_s} \angle(\delta - \theta)$$

$$= \left[\frac{V_t}{Z_s} \cos \theta - \frac{V_0}{Z_s} \cos(\delta - \theta)\right] + j\left[-\frac{V_t}{Z_s} \sin \theta - \frac{V_0}{Z_s} \sin(\delta - \theta)\right]$$

and $\mathbf{V}_0 = V_0 \cos \delta + jV_0 \sin \delta$. The developed power is given by the expression (in which * indicates the complex conjugate)

$$P_d = -\mathrm{Re}\,(\mathbf{V}_0\mathbf{I}_a^*) = -(V_0 \cos \delta)\left[\frac{V_t}{Z_s} \cos \theta - \frac{V_0}{Z_s} \cos (\delta - \theta)\right]$$

$$-\,(V_0 \sin \delta)\left[-\frac{V_t}{Z_s} \sin \theta - \frac{V_0}{Z_s} \sin (\delta - \theta)\right]$$

$$=\,-\frac{V_0 V_t}{Z_s} \cos (\delta + \theta) + \frac{V_0^2}{Z_s} \cos \theta$$

But $\cos \theta = R_a/Z_s$, and the desired result follows.

As will be shown in Problem 6.25 [see (1)], this result also obtains for generator action, when $P_d = +\mathrm{Re}(\mathbf{V}_0\mathbf{I}_a^*)$ and $\delta > 0$. In the limit as $R_a \to 0$, the expression for P_d coincides with (6.7), as it must.

6.18. A 2300-V, 3-phase, wye-connected, round-rotor synchronous motor has $X_s = 2\ \Omega$ per phase and $R_a = 0.1\ \Omega$ per phase. The motor operates at 0.866 leading power factor while taking a line current of 350 A. Find the rms value of the induced phase voltage and the power angle.

For motor operation,

$$\mathbf{V}_t = \mathbf{I}_a\mathbf{Z}_s + \mathbf{V}_0$$

Substituting

$$\mathbf{V}_t = \frac{2300}{\sqrt{3}}\ \angle 0° = 1328\angle 0°\ \mathrm{V}$$

$$\mathbf{I}_a = 350\angle 30°\ \mathrm{A}$$

$$\mathbf{Z}_s = 0.1 + j2 = 2\angle 87°\ \Omega$$

yields

$$1328\angle 0° = 700\angle 117° + \mathbf{V}_0 \quad\text{or}\quad 1328 = -318 + j624 + \mathbf{V}_0$$

whence $\mathbf{V}_0 = 1646 - j624 = 1760\angle -21°\ \mathrm{V}$. Thus, $V_0 = 1760\ \mathrm{V}$ and $\delta = -21°$.

6.19. A 2300-V, 3-phase, wye-connected, round-rotor synchronous motor has a synchronous reactance of $3\ \Omega$ per phase and an armature resistance of $0.25\ \Omega$ per phase. The motor operates on load such that the power angle is $-15°$, and the excitation is so adjusted that the internally induced voltage is equal in magnitude to the terminal voltage. Determine (a) the armature current and (b) the power factor of the motor.

(a)   From the phasor diagram, Fig. 6-18,

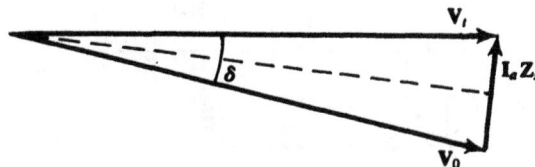

Fig. 6-18

$$I_a = \frac{2V}{Z_s} \left| \sin \frac{\delta}{2} \right|$$

where

$$V_0 = V_t = V = \frac{2300}{\sqrt{3}} = 1328 \text{ V}$$

$$Z_s = \sqrt{(3)^2 + (0.25)^2} = 3 \ \Omega$$

and $\delta = -15°$. Thus,

$$I_a = \frac{2(1328)}{3} \sin 7.5° = 115.6 \text{ A}$$

(b)   Neglecting R_a, we have

$$|P_d| = VI_a \cos \phi = \frac{V_0 V_t}{X_s} |\sin \delta|$$

or

$$\cos \phi = \frac{1328}{(115.6)(3)} \sin 15° = 0.991 \text{ lagging}$$

6.20. A 3-phase, wye-connected synchronous motor, rated at 15 hp, 400 V, has a full-load efficiency of 90%. Its synchronous reactance is 3 Ω per phase and the armature resistance is 0.15 Ω per phase. The motor operates at full-load and 0.8 leading power factor. Determine (a) the power angle and (b) the field current. The saturation characteristic is given in Fig. 6-19.

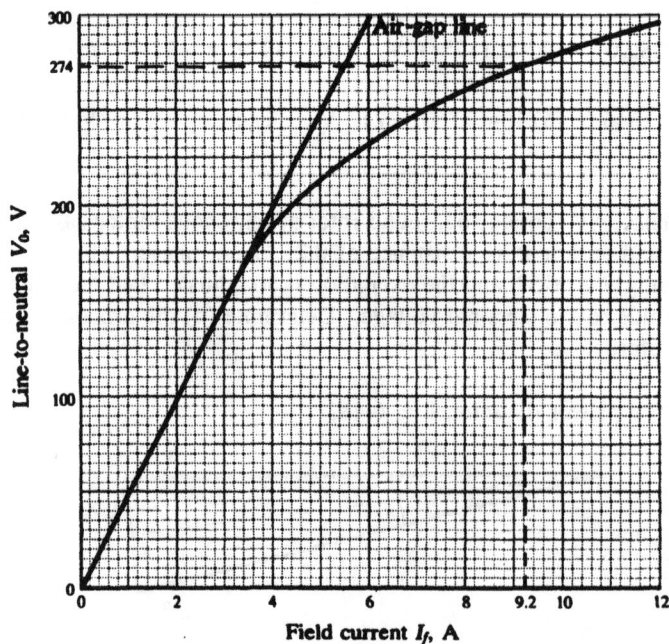

Fig. 6-19

(a) $\qquad$ phase voltage $V_t = \dfrac{400}{\sqrt{3}} = 231$ V

motor input $= \dfrac{15 \times 746}{0.90} = 12{,}433$ W

armature current $I_a = \dfrac{12{,}433/3}{(231)(0.8)} = 22.4$ A

$I_a Z_s = (22.4)\sqrt{(3)^2 + (0.15)^2} = 67.2$ V

The phasor diagram is shown in Fig. 6-20, where

$$\phi = \cos^{-1} 0.8 = 36.87° \qquad \theta = \tan^{-1}\dfrac{3}{0.15} = 87.13°$$

By the law of cosines,

$$V_0^2 = V_t^2 + (I_a Z_s)^2 - 2V_t(I_a Z_s)\cos(\theta + \phi)$$

$$= (231)^2 + (67.2)^2 - 2(231)(67.2)\cos 124° = 75\,237$$

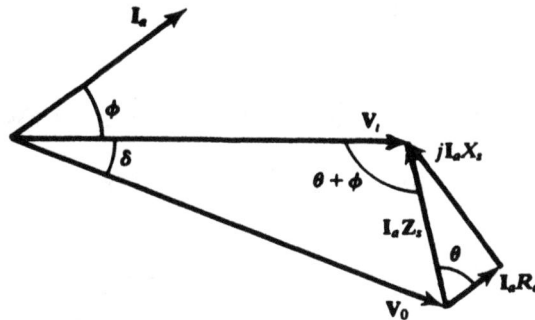

Fig. 6-20

or $V_0 = 274$ V. The, by the law of sines,

$$\left|\sin\delta\right| = \dfrac{I_a Z_s}{V_0}\sin(\theta + \phi) = \dfrac{67.2}{274}\sin 124° = 0.203$$

or $\delta = -11.7°$.

(b) $\qquad$ From Fig. 6-19, the field current corresponding to $V_0 = 274$ V is $I_f = 9.2$ A.

6.21. Calculate the minimum line current for the motor of Problem 6.20. Also, determine the corresponding field current. Neglect R_a.

The minimum armature current is

$$I_a = \dfrac{12433/3}{231} = 17.94$$ A

corresponding to $\cos\phi = 1$. From Fig. 6-20, with $\theta = 90°$ and $\phi = 0°$, we have

$$V_0^2 = V_t^2 + (I_a X_s)^2 = (231)^2 + (53.8)^2$$

which gives $V_0 = 237.2$ V. Then, from Fig. 6-19, $I_f = 6.2$ A.

6.22. The motor of Problem 6.20 runs with an excitation of 10 A and takes an armature current of 25 A. Neglecting the armature resistance, determine the developed power and the power factor.

From Fig. 6-19, at $I_f = 10$ A, $V_0 = 280$ V. From Fig. 6-20, since $\theta = 90°$,

$$V_0^2 = V_t^2 + (I_a X_s)^2 - 2V_t(I_a X_s) \cos (90° + \phi)$$

$$(280)^2 = (231)^2 + (75)^2 + 2(231)(75) \sin \phi$$

which gives $\cos \phi = 0.83$ leading. Then,

$$\text{developed power} = \text{input power} = \sqrt{3}\ (400)(25)(0.83) = 14.35\ \text{kW}$$

6.23. The motor of Problem 6.20 has 4 poles and is rated at 60 Hz. If I_f is adjusted such that $V_0 = V_t$, the motor takes 20 A in current. Neglecting R_a, determine the torque developed.

Proceeding as in Problem 6.22, we have

$$(231)^2 = (231)^2 + (60)^2 + 2(231)(60) \sin \phi$$

whence $\cos \phi = 0.99$ lagging.

$$\text{developed power} = \sqrt{3}\ (400)(20)(0.99) = 13.74\ \text{kW}$$

$$\text{synchronous speed} = \frac{120(60)}{4} = 1800\ \text{rpm}$$

$$\text{developed torque} = \frac{13740}{2\pi(1800)/60} = 72.9\ \text{N} \cdot \text{m}$$

6.24. An induction motor, while driving a load, takes 350 kW at 0.707 power factor lagging. An overexcited synchronous motor is then connected in parallel with the induction motor, taking 150 kW in power. If the overall power factor (of the two motors combined) becomes 0.9 lagging, calculate the kVA rating of the synchronous motor.

Induction motor:

$$\text{power} = 350\ \text{kW}$$

$$\text{apparent power} = \frac{350}{0.707} = 495\ \text{kVA}$$

$$\text{reactive power} = 495 \sin (-45°) = -350\ \text{kvar}$$

Total motor load:

$$\text{power} = 150 + 350 = 500\ \text{kW}$$

$$\text{apparent power} = \frac{500}{0.9} = 555.5\ \text{kVA}$$

$$\text{reactive power} = 555.5 \sin (\cos^{-1} 0.9) = -242.16\ \text{kvar}$$

Synchronous motor:

$$power = 150 \text{ kW}$$

$$reactive\ power = -242.16 - (-350) = 107.84 \text{ kvar}$$

$$apparent\ power = \sqrt{(150)^2 + (107.84)^2} = 184.74 \text{ kVA}$$

6.25. A round-rotor synchronous generator has a synchronous impedance Z_s per phase and a terminal voltage V_t per phase; it operates in parallel with an infinite bus (i.e., the voltage remains constant regardless of load fluctuations). The phasor diagram is shown in Fig. 6-21. Because of some disturbance, the power angle changes by ψ (as shown; $V_0' = V_0$), which causes the machine to develop an additional power, thereby keeping in synchronism. This additional power is known as the *synchronizing power*. Derive a general expression for the synchronizing power per phase and discuss the special cases when (a) ψ is very small, (b) ψ is very small and $R_a \ll X_s$.

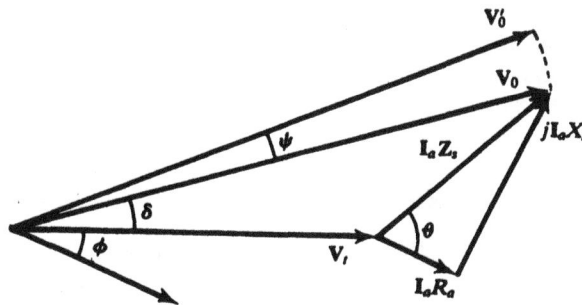

Fig. 6-21

From the phasor diagram,

$$\mathbf{I}_a = \frac{V_0 \angle \delta - V_t \angle 0°}{Z_s \angle \theta} = \frac{1}{Z_s} \left[V_0 \angle (\delta - \theta) - V_t \angle -\theta \right]$$

$$= \frac{1}{Z_s} \left[V_0 \cos (\delta - \theta) - V_t \cos \theta \right] + \frac{j}{Z_s} \left[V_0 \sin (\delta - \theta) + V_t \sin \theta \right]$$

and $\mathbf{V}_0 = V_0 \cos \delta + j V_0 \sin \delta$. The power internally developed is then given by (an * denotes the complex conjugate)

$$P_d = \text{Re} \,(\mathbf{V}_0 \mathbf{I}_a^*) = \frac{V_0}{Z_s} \cos \delta \left[V_0 \cos (\delta - \theta) - V_t \cos \theta \right]$$

$$+ \frac{V_0}{Z_s} \sin \delta \left[V_0 \sin (\delta - \theta) + V_t \sin \theta \right] \tag{1}$$

$$= \frac{V_0}{Z_s} \left[V_0 \cos \theta - V_t \cos(\theta + \delta) \right]$$

When δ becomes $\delta + \psi$, we get, from (1), the new internal power, P_d':

$$P_d' = \frac{V_0}{Z_s} \left[V_0 \cos \theta - V_t \cos (\theta + \delta + \psi) \right] \tag{2}$$

Subtracting (1) from (2), we have for the synchronizing power

$$P_s = P_d' - P_d = \frac{V_0 V_t}{Z_s}\left[\sin\psi\sin(\theta+\delta) + 2\cos(\theta+\delta)\sin^2\frac{\psi}{2}\right] \qquad (3)$$

which is the general result.

(a)   When ψ (in rad) is very small such that $\sin^2(\psi/2) \to 0$ and $\sin\psi \to \psi$, then

$$P_s = \frac{V_0 V_t}{Z_s}\,\psi\,\sin(\theta+\delta)$$

(b)   If, furthermore, $R_a \ll X_s$, then $\theta \approx 90°$ and

$$P_s = \frac{V_0 V_t}{X_s}\,\psi\,\cos\delta$$

6.26. A 1000-kVA, 3300-V, 60-Hz, 4-pole, 3-phase, wye-connected, round-rotor synchronous generator has a synchronous reactance of 1 Ω per phase; the armature resistance is negligible. The generator is operating at full-load, 0.8 lagging power factor on an infinitie bus. Calculate the synchronizing power per phase and synchronizing torque per phase, if a disturbance causes the power angle to swing by one (mechanical) degree.

The relationship between electrical and mechanical angular measures is

θ (electrical degrees) = [θ (mechanical degrees)] × (number of pole pairs)

Thus, the change in the power angle is

$$\psi = 1 \times \frac{4}{2} = 2° \text{ electrical}$$

Now, $\mathbf{V}_0 = \mathbf{V}_t + j\mathbf{I}_a X_s$. Choose $\mathbf{I}_a$ as the reference phasor. Then:

$$\mathbf{V}_t = \frac{3300}{\sqrt{3}}(0.8 + j0.6) = 1524 + j1143 \quad \text{V}$$

$$\mathbf{I}_a = \frac{1000}{\sqrt{3}\,(3.3)} = 175 + j0 \quad \text{A}$$

$$j\mathbf{I}_a X_s = 0 + j175 \quad \text{V}$$

$$\mathbf{V}_0 = 1524 + j1318 = 2015\angle 41°$$

$$\phi = \cos^{-1} 0.8 = 37°$$

$$\delta = 41° - 37° = 4°$$

Substituting in (3) of Problem 6.25,

$$P_s = \frac{(2015)(3300/\sqrt{3})}{1}\left[\sin 2°\sin(90°+4°) + 2\cos(90°+4°)\sin^2 1°\right] = 133.6 \text{ kW}$$

and

$$T_s = \frac{133600}{2\pi(1800)/60} = 709 \text{ N}\cdot\text{m}$$

6.27. Two identical 3-phase, wye-connected synchronous generators share equally a load of 10 MW at 33 kV and 0.8 lagging power factor. The synchronous reactance of each machine is 6 Ω per phase and the armature resistance is negligible. If one of the machine has its field excitation adjusted to carry 125 A lagging current, what is the current supplied by the second machine? The prime mover inputs to both machines are equal.

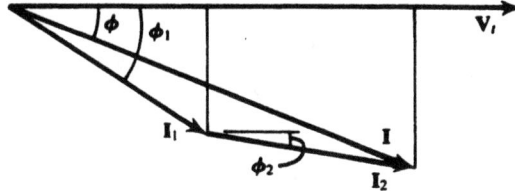

Fig. 6-22

The phasor diagram of current division is shown in Fig. 6-22, wherein $I_1 = 125$ A. Because the machines are identical and the prime mover inputs to both machines are equal, each machine supplies the same true power:

$$I_1 \cos \phi_1 = I_2 \cos \phi_2 = \frac{1}{2} I \cos \phi$$

Now

$$I = \frac{10 \times 10^6}{\sqrt{3} \ (33 \times 10^3)(0.8)} = 218.7 \text{ A}$$

whence

$$I_1 \cos \phi_1 = I_2 \cos \phi_2 = \frac{1}{2} (218.7)(0.8) = 87.5 \text{ A}$$

The reactive current of the first machine is therefore

$$I_1 \left| \sin \phi_1 \right| = \sqrt{(125)^2 - (87.5)^2} = 89.3 \text{ A}$$

and since the total reactive current is

$$I \left| \sin \phi \right| = (218.7)(0.6) = 131.2 \text{ A}$$

the reactive current of the second machine is

$$I_2 \left| \sin \phi_2 \right| = 131.2 - 89.3 = 41.9 \text{ A}$$

Hence

$$I_2 = \sqrt{(87.5)^2 + (41.9)^2} = 97 \text{ A}$$

6.28. Consider the two machines of Problem 6.27. If the power factor of the first machine is 0.9 lagging and the load is shared equally by the two machines, what are the power factor and current of the second machine?

Load: power = 10 000 kW
                       apparent power = 12 500 kVW
                       reactive power = −7500 kvar
First machine: power = 500 kW
                       $\phi_1 = \cos^{-1} 0.9 = -25.8°$
                       reactive power = 5000 tan ϕ_1 = −2422 kva

Second machine: power = 5000 kW

reactive power = −7500 − (−2422) = −5078 kvar

$$\tan \phi_2 = \frac{-5078}{5000} = -1.02$$

$$\cos \phi_2 = 0.7$$

$$I_2 = \frac{5000}{\sqrt{3}\,(33)(0.7)} = 124.7 \text{ A}$$

6.29. A synchronous motor has moment of inertia 20 kg · m² and damping coefficient 15 N · m · s/rad. If the machine has a synchronizing torque constant of 800 N · m/rad, determine the natural frequence of oscillation and the damping ratio for small disturbances.

From (6.20), if $\psi \equiv \Delta\delta$ is a small variation in the power angle,

$$J\ddot{\psi} + b\dot{\psi} + k_e\psi = 0$$

Then, by (6.21),

$$f_n = \frac{1}{2\pi}\sqrt{\frac{k_e}{J}} = \frac{1}{2\pi}\sqrt{\frac{800}{20}} \approx 1 \text{ Hz}$$

$$\zeta = \frac{b}{2\sqrt{k_e J}} = \frac{15}{2\,\sqrt{(800)(20)}} = 0.06$$

6.30. A 3-phase short-circuit test is performed on a synchronous generator for which the envelope of the armature current is shown in Fig. 6-13. Given $V_0 = 231$ V; $\Delta i'' = 113$ A; $\Delta i' = 117$ A; the steady-state short-circuit current is 144 A. Determine (a) X_d, (b) x'_d and (c) x''_d.

(a)
$$X_d = \frac{V_0}{i_d} = \frac{231}{144} = 1.6 \ \Omega$$

(b)
$$\Delta i' = 0.368(i'_d - i_d)$$

$$117 = 0.368(i'_d - 144)$$

$$i'_d = 462 \text{ A}$$

and

$$x'_d = \frac{V_0}{i'_d} = \frac{231}{462} = 0.5 \ \Omega$$

(c)
$$\Delta i'' = 0.368(i''_d - i'_d)$$

$$113 = 0.368(i''_d - 462)$$

$$i''_d = 769 \text{ A}$$

and

$$x_d'' = \frac{V_0}{i_d''} = \frac{231}{769} = 0.3 \ \Omega$$

6.31. The machine of Problem 6.30 is rated at 100 kVA, 400 V, and is wye-connected. Express X_d, x_d' and x_d'' in per unit.

$$\text{base apparent power} = 100 \ \text{kVA}$$

$$\text{base phase voltage} = \frac{400}{\sqrt{3}} = 231 \ \text{V}$$

$$\text{base current} = \frac{100 \times 10^3}{\sqrt{3}\,(400)} = 144.3 \ \text{A}$$

$$\text{base impedance} = \frac{231}{144.3} = 1.6 \ \Omega$$

Consequently,

$$\text{per-unit } X_d = \frac{1.6}{1.6} = 1 \ \text{pu}$$

$$\text{per-unit } x_d' = \frac{0.5}{1.6} = 0.31 \ \text{pu}$$

$$\text{per-unit } x_d'' = \frac{0.3}{1.6} = 0.19 \ \text{pu}$$

Supplementary Problems

6.32. What is the maximum speed at which (a) a 60-Hz, (b) a 50-Hz, synchronous machine can be operated? *Ans.* (a) 3600 rpm; (b) 3000 rpm

6.33. The flux-density distribution in the airgap of a 1200-rpm, 60-Hz synchronous machine is $B(\theta) = 0.7 \cos \theta$. The armature has an 80-turn coil in each phase and is wye-connected. Determine the rms value of the line voltage if the axial length of the armature is 160 mm and its diameter is 120 mm. *Ans.* 165.5 V

6.34. The machine of Problem 6.33 has 108 slots, with 4 conductors per slot, and has a distributed winding with a coil pitch of 16 slots. Determine the rms phase voltage. *Ans.* 74.8 V

6.35. Calculate the distribution factor, pitch factor, and winding factor for the machine of Problem 6.34. *Ans.* 0.9561; 0.9848; 0.9416

6.36. If the machine of Problem 6.34 has 10% third harmonic present in the airgap flux density, and the armature winding is full-pitched, determine the rms value of the phase voltage.   *Ans.* 98.84 V

6.37. A 60-turn coil mounted on an armature 120 mm in diameter and 100 mm in axial length rotates at 3000 rpm in a uniform magnetic field of 0.5 T. What is the instantaneous voltage induced in the coil? *Ans.* 113 sin 314t (V)

6.38. Draw a sketch of a salient-pole synchronous machine. Define the direct- and quadrature-axis synchronous reactances. Which one is the larger? Why?

6.39. A 25-kVA, 3-phase, wye-connected, 400-V synchronous generator has a synchronous impedance of 0.05 + j1.6 Ω per phase. Determine the full-load voltage regulation at (*a*) 0.8 power factor lagging, (*b*) unity power factor, (*c*) 0.8 power factor leading.   *Ans.* (*a*) 22.2%; (*b*) 10.6%; (*c*) −5.5%

6.40. Determine the power angles for the three cases in Problem 6.39.
Ans. (*a*) 7.2°; (*b*) 13°; (*c*) 15°

6.41. The generator of Problem 6.39 is to have zero voltage regulation at half full-load. Neglecting the armature resistance, find the operating power factor and the developed power.
Ans. 0.997 leading; 12.5 kW

6.42. A 500-kVA, 6-pole, 500-V, 3-phase, wye-connected synchronous generator has a synchronous impedance of 0.1 + j1.5 Ω per phase. If the generator is driven at 1000 rpm, what is the frequency of the generated voltage? Determine the excitation voltage and the power angle on full-load and 0.8 lagging power factor. *Ans.* 50 Hz; 1078 V; 37.6°

6.43. A 100-kVA, 400-V, wye-connected, salient-pole synchronous generator runs at full-load and 0.8 leading power factor. If X_d = 2X_q = 1.1 Ω per phase and R_a is negligible, calculate (*a*) the voltage regulation, (*b*) the power angle, and (*c*) the developed power.   *Ans.* (*a*) −26.4%; (*b*) 19°; (*c*) 80 kW

6.44. From the data of Fig. 6-15, plot X_s, against I_f.

6.45. A 30-kVA, 3-phase, 230-V, wye-connected synchronous generator has a synchronous reactance of 0.8 Ω per phase. The armature resistance is negligible. Calculate the percent voltage regulation on (*a*) full-load at 0.8 power factor leading, (*b*) 50% full-load at unity power factor, (*c*) 25% full-load at 0.8 power factor lagging.   *Ans.* (*a*) −18.7%; (*b*) 2.5%; (*c*) 7.2%

6.46. For each of the three cases of Problem 6.45, compute the power angle.
Ans. (*a*) 26.5°; (*b*) 12.8°; (*c*) 4.8°

6.47. Verify from (*6.12*), with $X_q = X_s$, that the power angle obtained in Problem 6.18 is correct.

6.48. Refer to Problem 6.17. (*a*) For what power angle is maximum electrical power taken by the motor? (*b*) What is the magnitude of this maximum power (per phase)? (*c*) What is the corresponding excitation limit?

Ans. (*a*) $\delta = -\theta$; (*b*) $|P_m| = \dfrac{V_0 V_t}{Z_s} - \dfrac{V_0^2 R_a}{Z_s^2}$; (*c*) $V_0 = \dfrac{Z_s}{2R_a}(V_t - \sqrt{V_t^2 - 4R_a\,|P_m|})$

6.49.   A 3-phase, wye-connected, round-rotor synchronous generator has $X_s = 1.2\ \Omega$ per phase and $R_a = 0.4\ \Omega$ per phase. The generator supplies a load of 30 kVA at 220 V and 0.8 power factor lagging. The excitation characteristic is shown in Fig. 6-19. Calculate the power angle and the field current.
Ans. 15°; 5.2 A

6.50.   A 3-phase, Y-connected, 220-V (line-to-line) synchronous generator supplies a capacitive load of $5\ \angle{-30°}$ Ω per phase. The per-phase synchronous reactance is $j5$ ohm, and the armature resistance is negligible. Calculate (*a*) the voltage regulation; (*b*) power angle, and (*c*) the apparent power supplied by the generator at the above load.   *Ans.* (*a*) 0; (*b*) 60°; (*c*) 9678.7 VA

6.51.   The machine of Problem 6.50 is run as an overexcited synchronous motor while developing maximum power. If the excitation voltage (V_0) is 127 V/phase, calculate the armature current.   *Ans.* 35.92 A

6.52.   A synchronous motor driving a fan is operating at unity power factor. Discuss in a qualitative manner (using phasor diagrams if helpful) the effect upon motor power factor, armature current, power angle, and speed (*a*) when the field current is increased 10% and the terminal voltage is held constant, (*b*) when the terminal voltage is increased 10% and the field current is held constant.
Ans. (*a*) power factor leading, I_a increases, δ decreases, no change in speed; (*b*) power factor lagging, small decrease in I_a, δ decreases, no change in speed

6.53.   A synchronous motor is operating at half full-load. An increase in its field current causes a decrease in its armature current. Before the change in field current, did the armature current lead or lag the terminal voltage? Justify your answer.   *Ans.* lag

6.54.   A synchronous motor is operating at rated load and unity power factor. The field current is increased 20%. Show the resulting changes in all voltage and current phasors on a phasor diagram.

6.55.   A 400-V, 3-phase, wye-connected, round-rotor synchronous motor operates at unity power factor while developing a power of 60 kW. If the synchronous reactance is $1.0\ \Omega$ per phase and the armature resistance is negligible, calculate (*a*) the induced voltage per phase and (*b*) the power angle.
Ans. (*a*) 246.6 V; (*b*) −20.5°

6.56.   An overexcited 2300-V, 3-phase, wye-connected synchronous motor runs at a power angle of −21°. The per-phase synchronous impedance is $0.1 + j2\ \Omega$. If the motor takes a line current of 350 A, determine the power factor.   *Ans.* 0.87 leading

6.57.   What are (*a*) the power factor and (*b*) the line current of the motor of Problem 6.56, if the internal induced voltage is the same as the line voltage and the power angle is −20°? (*c*) Also determine the developed power.   *Ans.* (*a*) 0.99 lagging; (*b*) 242 A; (*c*) 938 kW

6.58.   A 400-V, 3-phase, round-rotor synchronous motor has an efficiency of 92% while delivering 18 hp (at the shaft). The per-phase synchronous impedance is $0.5 + j1.5\ \Omega$. If the motor operates at 0.9 lagging power factor, determine (*a*) the power angle and (*b*) the field current. The motor saturation characteristic is shown in Fig. 6-19.   *Ans.* (*a*) −7.4°; (*b*) 4.5 A

6.59.   The motor of Problem 6.58 has 6 poles and is rated at 60 Hz. If the field excitation is 8 A and the motor takes 20 A in current, calculate the developed torque.   *Ans.* 50 N · m

6.60. An overexcited synchronous motor is connected across a 100-kVA inductive load having a 0.8 lagging power factor. The motor takes 10 kW in power while idling (on no-load). Calculate the kVA rating of the motor if it is desired to bring the overall power factor to unity. The motor is not used to carry any load.   *Ans.* 60.8 kVA

6.61. A round-rotor synchronous generator operates on an infinite bus at 2300 V across the lines. The generator is wye-connected, has a synchronous reactance of 2 Ω per phase and negligible armature resistance, and supplies a current of 300 A at 0.8 lagging power factor. A disturbance causes the power angle to swing 2° electrical. Determine the synchronizing power per phase.   *Ans.* 122 kW

6.62. Two wye-connected, identical synchronous generators, operating in parallel, share equally a 1-MW load at 11 kV and 0.8 lagging power factor. If one of the machines supplies 40 A at a lagging power factor, determine (*a*) the current and (*b*) the power factor of the second machine.
Ans. (*a*) 27.8 A; (*b*) 0.94 lagging

6.63. Obtain the power factor and induced voltage for the second generator of Problem 6.27.
Ans. 0.9 lagging; 19.3 kV

6.64. A synchronous motor is delivering 50 N $\cdot$ m in torque at 3600 rpm. The load torque is suddenly reduced to zero and the power angle is observed to oscillate initially over a 12° (electrical) range with a period of 0.1 s. After 6 s, the oscillations have decreased to 4°. Calculate the synchronizing torque constant, J and b for the zero-load condition.   *Ans.* $k_e = 239$ N $\cdot$ m/rad; $J = 0.06$ kg $\cdot$ m^2; $b = 0.00817$ N $\cdot$ m $\cdot$ s/rad

6.65. A 3-phase 800 kVA 11 kV (line-to-line) wye-connected synchronous generator has $R_a = 1.5$ ohm/phase and $X_s = 25$ ohm/phase. For a 600 kW 0.8 leading power factor load, calculate the (*a*) percent voltage regulation; (*b*) power angle; and (*c*) total internal (or developed) power.
Ans. (*a*) −7.13%; (*b*) 7.4°; (*c*) 607 kW

6.66. A 3-phase 400 V (line-to-line) wye-connected synchronous motor has $X_s = 1.0$ ohm/phase and a negligible armature resistance. The field current is so adjusted that the internal (induced) voltage is 270 V/phase, while the motor draws 40 kW power from the source. Calculate the (*a*) power factor of the motor; and (*b*) reactive power supplied by the motor to the source.   *Ans.* (*a*) 0.87 leading; (*b*) −22.72 kvar

Chapter 7

Single-Phase Motors
and Permanent Magnet Machines

7.1 SMALL AC MOTORS

Almost invariably, small ac motors are single-phase motors. As such, they are not self-starting, because the magnetic field produced by a single-phase winding is a pulsating field rather than a rotating field. To make the motor self-starting, it is provided with an auxiliary starting winding which causes it to act like an unbalanced two-phase machine during starting. (See Section 7.3).

In this chapter we confine our attention to single-phase induction motors and hysteresis motors, which are small motors of the synchronous type, such as are used in clocks and turntables. We briefly discuss permanent magnet machines also.

7.2 ANALYSIS OF SINGLE-PHASE INDUCTION MOTORS

An approach to single-phase induction motors is based on the fact that a pulsating field,

$$B(\theta, \cdot t) = B_m \cos k\theta \sin \omega t \qquad (7.1)$$

such as is produced by the main winding of the motor, may also be expressed as

$$B(\theta, t) = \frac{B_m}{2} \sin (\omega t - k\theta) + \frac{B_m}{2} \sin (\omega t + k\theta) \qquad (7.2)$$

Thus, we have two counter rotating fields, and the theory based on this concept is known as the *double-revolving field theory*. The forward direction of rotation is defined as the direction of rotation of the rotor. Thus, we may define a slip s_f of the rotor with respect to the forward-rotating field as

$$s_f \equiv s = \frac{n_s - n}{n_s} \qquad (7.3)$$

where $n_s = (60\omega/2\pi k)$ (rpm). Notice that s_f is similar to the slip s of the polyphase induction motor. We may also define a slip s_b of the rotor with respect to the backward-rotating field as

$$s_b = \frac{-n_s - n}{-n_s} = 2 - s \qquad (7.4)$$

The torque relationship of the polyphase induction motor is applicable to each of the two rotating fields of the single-phase motor, except that the amplitude of each rotating field is one-half that of the alternating field. This results in an equal division of magnetizing and leakage reactances, and the approximate equivalent circuit becomes as shown in Fig. 7-1(*a*). The torque-speed characteristic is shown in Fig. 7-1(*b*). Having developed the equivalent circuit, the performance calculations for the single-phase motor are very similar to those for the polyphase motor. See Problems 7.1 through 7.7.

7.3 STARTING OF SINGLE-PHASE INDUCTION MOTORS

We already know that because of the absence of a rotating magnetic field, when the rotor of a single-phase induction motor is at standstill, it is not self-starting. The two methods of starting a single-phase motor

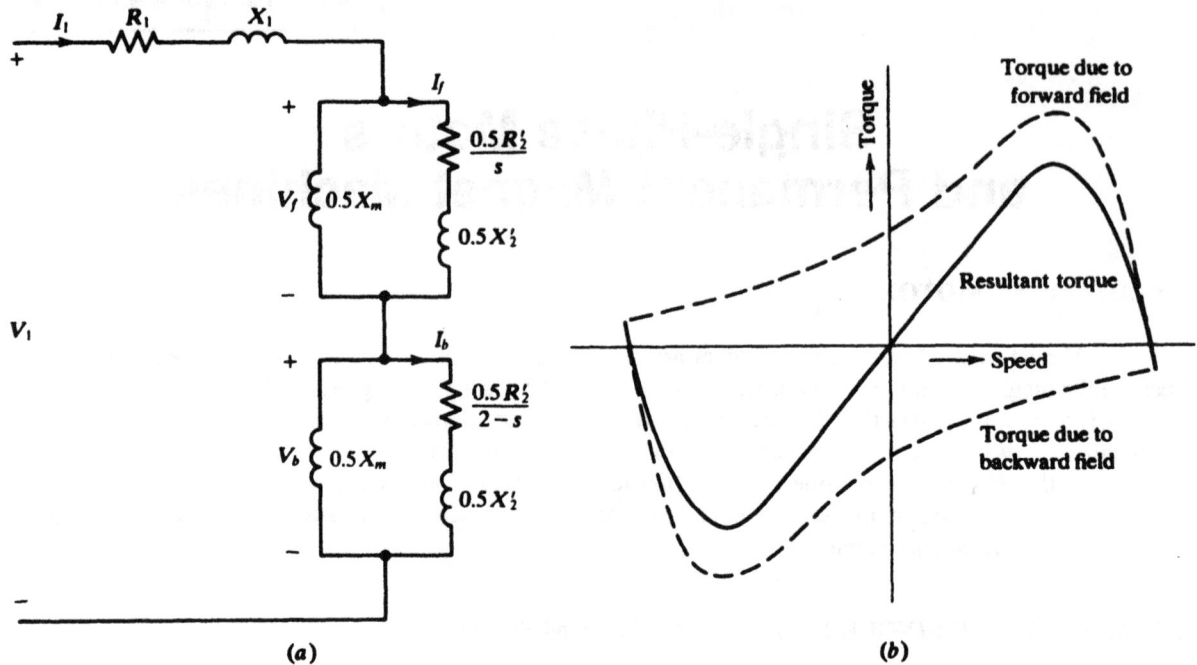

Fig. 7-1

are either to introduce commutator and brushes, such as in a repulsion motor, or to produce a rotating field by means of an auxiliary winding, such as by split phasing. We consider the latter method here.

From the theory of the polyphase induction motor, we know that in order to have a rotating magnetic field, we must have at least two mmf's which are displaced from each other in space and carry currents having different time phases. Thus, in a single-phase motor, a starting winding on the stator is provided as a source of the second mmf. The first mmf arises from the main stator winding. The various methods to achieve the time and space phase shifts between the main winding and starting winding mmf's are summarized below.

Split-Phase Motors

This type of motor is represented schematically in Fig. 7-2(a), where the main winding has a relatively low resistance and a high reactance. The starting winding, however, has a high resistance and a low reactance, and a centrifugal switch as shown. The phase angle α between the two currents I_m and I_s is about 30 to 45°, and the starting torque T_s is given by

$$T_s = K I_m I_s \sin \alpha \qquad (7.5)$$

where K is a constant. When the rotor reaches a certain speed (about 75 percent of its final speed), the centrifugal switch comes into action and disconnects the starting winding from the circuit. The torque-speed characteristic of the split-phase motor is of the form shown in Fig. 7-2(b). Such motors find applications in fans, blowers, and so forth, and are rated up to ½ hp.

A higher starting torque can be developed by a split-phase motor by inserting a series resistor in the starting winding. A somewhat similar effect may be obtained by inserting a series inductive reactance in the main winding. This reactance is short-circuited when the motor builds up speed.

Capacitor-Start Motors

By connecting a capacitor in series with the starting winding, as shown in Fig. 7-3, the angle α in (7.5) can be increased. The motor will develop a higher starting torque by doing this. Such motors are not restricted merely to fractional-horsepower ratings, and may be rated up to 10 hp. At 110 V, a 1-hp motor requires a capacitance of about 400 µF, whereas 70 µF is sufficient for a ⅛-hp motor. The capacitors generally used are inexpensive electrolytic types and can provide a starting torque that is almost four times that of the rated torque.

As shown in Fig. 7-3, the capacitor is merely an aid to starting and is disconnected by the centrifugal switch when the motor reaches a predetermined speed. However, some motors do not have the centrifugal switch. In such a motor, the starting winding and the capacitor are meant for permanent operation and the capacitors are much smaller. For example, a 110-V ½-hp motor requires a 15-µF capacitance.

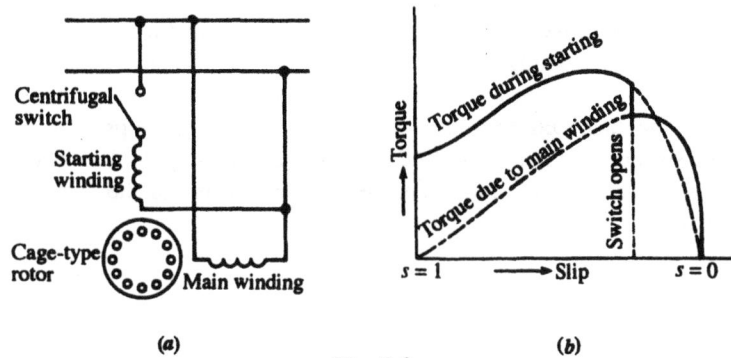

Fig. 7-2

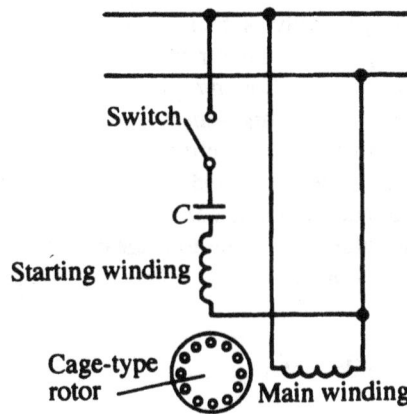

Fig. 7-3

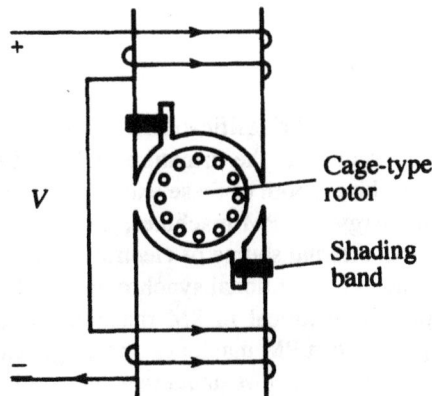

Fig. 7-4

A third kind of capacitor motor uses two capacitors: one that is left permanently in the circuit together with the starting winding, and one that gets disconnected by a centrifugal switch. Such motors are, in effect, unbalanced two-phase induction motors.

Shaded-Pole Motors

Another method of starting very small single-phase induction motors is to use a shading band on the poles, as shown in Fig. 7-4, where the main single-phase winding is also wound on the salient poles. The shading band is simply a short-circuited copper strap wound on a portion of the pole. Such a motor is known as a shaded-pole motor. The purpose of the shading band is to retard (in time) the portion of flux passing through it in relation to the flux coming out of the rest of the pole face. Thus the flux in the unshaded portion reaches its maximum before that located in the shaded portion. And we have a progressive shift of flux from the direction of the unshaded portion to the shaded portion of the pole, as shown in Fig. 7-4. The effect of the progressive shift of flux is similar to that of a rotating flux, and because of it, the shading band provides a starting torque. Shaded-pole motors are the least expensive of the fractional-horsepower motors and are generally rated up to 1/20 hp.

7.4 PERMANENT MAGNET MACHINES

Permanent magnet (PM) machines compose a well-known class of machines used in both the motoring and generating modes. PM machines have been used for many years in applications where simplicity of structure and a low initial cost were of primary importance. PM machines have also been applied to more demanding applications, primarily as the result of the availability of low-cost power electronic control devices and the improvement of permanent magnet characteristics. In general, modern PM machines are competitive both in performance and cost with many types of field-wound dc machines and single-phase synchronous machines; the combinations of high-energy permanent magnets and solid-state power semiconductors are the principal ingredients of a relatively new class of machines commonly known as "brushless dc machines," and more appropriately termed "self-synchronous machines."

Energy converters using permanent magnets come in a variety of configurations and are described by such terms as motor, generator, alternator, stepper motor, linear motor, actuator, transducer, control motor, tachometer, brushless dc motor, and many others. In the following discussion, most of the presentations will refer to

1. Conventional (commutator) dc motor/generator
2. Synchronous alternator
3. Brushless dc motor
4. Digital machines

Unique Features of PM Machines

PM machines fall into a generalized classification known as "doubly excited" machines, which have two sources of excitation—usually known as the armature and the field (or excitation). In conventional synchronous and dc-commutator machines, both of these excitation sources are electrical windings connected to an external source of electrical energy. In PM machines, the excitation or field winding is replaced by a permanent magnet and, of course, no external source of electrical energy is required. In other respects, a PM machine may be directly comparable to conventional synchronous or dc-commutator machines, and armature windings and magnetic circuit may be identical in PM machines as compared to conventional machines. There is no comparison or analogy between PM machines and singly excited machines such as the induction motor or hysteresis machines. However, PM machines generally have the structural simplicity of singly-excited machines and are, therefore, often compared with singly excited machines in terms of cost, ease of assembly, size, and volume.

Permanent Magnet Materials

In Section 1.9 (Chapter 1) we alluded to permanent magnets and some of their characteristics. We now consider these topics in more detail.

A great many types of PM materials have been developed in the twentieth century, particularly in the latter half of this century, and there are many trends and indications that new PM materials will continue to be developed in the years ahead. As a result, the PM machine designer has a large choice of PM materials for consideration in most designs and often has the ability to optimize a design for minimum size, weight, cost, or other design specification by means of PM material selection. Most commercially used magnets fall within six general categories, which help to reduce the characteristics of which one should have at least a slight cognizance. Within each of these categories, there is still a myriad of variations in many parameters, but these are primarily due to the differences among materials manufacturers and to manufacturing techniques. Therefore, when precise magnetic parameters are required in a specific application, precise specifications should be made by the designer and agreed upon by the manufacturer.

Alnico Magnets

Alnico PM materials are metallic alloys of aluminum, nickel, cobalt, and iron, and were among the first high-energy PMs to be developed. Alnico magnets are, generally, characterized by relatively high residual flux density (B_r) and relatively low coercive force (H_c). The latter characteristic is undesirable from the electric machine standpoint. Certain grades of Alnico, such as grade 8HC, have been developed to remedy this weakness, but at the expense of lowered residual density. The characteristics of several popular grades of Alnico are summarized in Fig. 1-11 (Chapter 1). Alnico magnets are manufactured in "generic grades" from 1 through 9, with many variations within each grade. The grades generally represent the chronological order in which the properties designated by these grades were developed commercially.

Ceramic Magnets

Ceramic magnets are similar to other types of materials we commonly refer to as ceramics in physical properties, and, hence, the popular name. However, ceramic PMs are properly defined as ferrite oxides of barium or strontium and exhibit the property known as ferromagnetism. Due to the types of materials and manufacturing processes used, ceramic magnets are generally the lowest cost magnets available in terms of cost per unit of energy product. Ceramic magnets are by far the most widely used of any PM types in almost all applications, including rotating machines.

Ceramic magnets are characterized by relatively low residual flux densities (B_r) and relatively high coercive forces (H_c). Because of the latter characteristic, ceramic magnets are able to withstand armature reaction fields without demagnetization and are well suited for electric machine applications. Although ceramic magnets have generally poor mechanical and structural characteristics, they are the lightest in density of the common magnet types. This is often a distinct advantage in machine applications and tends to compensate for the increased pole-face area required due to the low residual flux density. Also, ceramic magnets have the lowest recoil permeability of common magnets, which is a stabilizing factor in machine application.

Samarium Cobalt Magnets

Samarium-cobalt magnets gave an "order-of-magnitude leap" in energy product over ceramic magnets and most other types of magnets. Samarium-cobalt magnets have residual flux densities comparable with Alnicos and coercive forces three to five times those of ceramic magnets. These magnets also have generally improved physical characteristics as compared to both Alnicos and ceramics. From a technical standpoint, they are ideal for rotating electric machine applications.

Demagnetization characteristics of certain samarium-cobalt magnets are shown in Fig. 7-5.

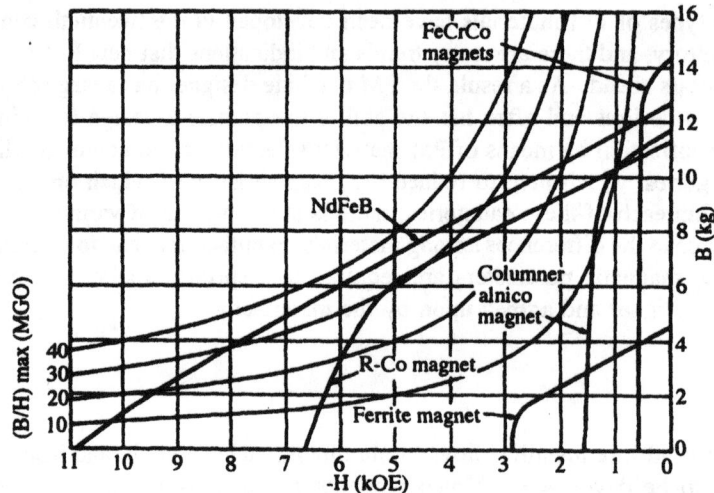

Fig. 7-5

Neodymium-iron-boron (NdFeB) Magnets

Neodymium-iron-boron (NdFeB) PM materials appear to offer the greatest promise for a PM material with greatly improved characteristics over those of ceramic magnets. This material has been shown in the laboratory to have the highest energy product of any PM material, and commercial versions of these laboratory samples are available with energy products above those of samarium cobalt. Perhaps more importantly, NdFeB magnets hold the promise of relatively low cost in production quantities.

NdFeB has the highest coercive force available in commercial magnets and, therefore, is ideally suited for machine applications. Also, its residual flux density is relatively high, comparable to the best of the Alnicos. As stated above, its energy product is the highest available today. Limitations of this material include very poor temperature characteristics. The low operating temperature requires the use of a larger size for an application required to operate at elevated temperatures and, therefore, many of the reduced size and weight advantages of NdFeB are lost.

A trade name of NdFeB magnets is Magnaquench. Its characteristics are shown in Fig. 7-6.

Permanent Magnet DC Machines

The armature of a permanent magnet dc machine is very similar to that of a field-excited dc machine discussed in Chapter 4. However, the field excitation in a PM dc machine is provided by permanent magnets. In this respect, a PM dc machine is similar to a conventional shunt machine with a constant field excitation. Thus, the governing equations of a PM dc machine become:

$$T_d = \frac{ZP}{2\pi a}(\phi I_a) = K\phi I_a \tag{7.6}$$

$$E = K\phi \omega_m \tag{7.7}$$

$$V = E \pm I_a R_a \tag{7.8}$$

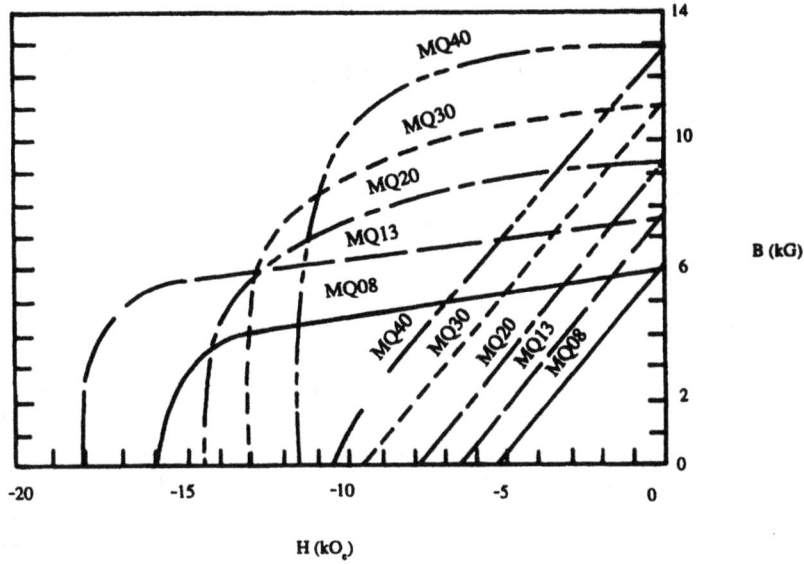

Fig. 7-6

$$\phi = \frac{\pi D l B_g}{P} \qquad (7.9)$$

where T_d = developed torque, N-m
   E = developed emf, V
   I_a = armature current, A
   ϕ = flux per pole, Wb
   V = terminal voltage, V
   R_a = armature (including brush) resistance, ohm
   ω_m = armature rotational speed, rad/s
   D = stator bore, m
   l = stator stack length, m
   K = winding factor = $ZP/(2\pi a)$
   Z = total armature conductors
   P = number of poles
   a = number of armature parallel paths

Combining (7.6) and (7.9) gives

$$T_d = \frac{Z(D l B_g I_a)}{2a} \qquad (7.10)$$

Letting $Z I_a/(2a) = X\pi D$ gives

$$T_d = \pi D^2 l B_g X = 4 \, (\text{vol}) B_g X \qquad (7.11)$$

where X = electrical loading in armature, A/m, and vol = magnetic volume = $\pi D^2 l/4$.
   For a specified developed torque and electrical loading, (7.11) can be expressed as:

$$(\text{vol}) \, B_g = \frac{T_d}{X} = \text{constant} \qquad (7.12)$$

Equation (*7.12*) is valid for any type of dc commutator machine for which the preceding equations are valid, but it is particularly useful in the evaluation and design of PM machines in which B_g is a function of the type of PM used for excitation. Note that (*7.12*) ignores thermal limitations of a machine, which often may be the primary limitation on reducing the volume of a machine.

If follows from (*7.6*) that the torque/speed characteristics of PM dc motor would be linear.

Permanent Magnet Synchronous Machines

The vast array of synchronous machine configurations in the medium and low-power ranges can generally be classified into two groups: *conventional* and *brushless*. PM machines fall into the latter group. PM synchronous machines generally have the same operating and performance characteristics as synchronous machines in general: operation at synchronous speed, a single or polyphase source of ac supplying the armature windings, a power limit above which operation at synchronous speed is unstable, reversible power flow, damper (cage) windings for starting and stability purposes, a torque angle between armature and field phasors, etc. A PM machine can have a configuration almost identical to that of the conventional synchronous machine with the absence of slip rings and a field winding. This absence, of course, is responsible for the one major difference between a PM machine and a conventional synchronous machine: lack of power factor or reactive power control and its association with terminal voltage regulation.

Fig. 7-7 shows the cross-section of a very simple PM synchronous machine. With the magnets mounted on the rotor, the machine could work as a brushless dc motor.

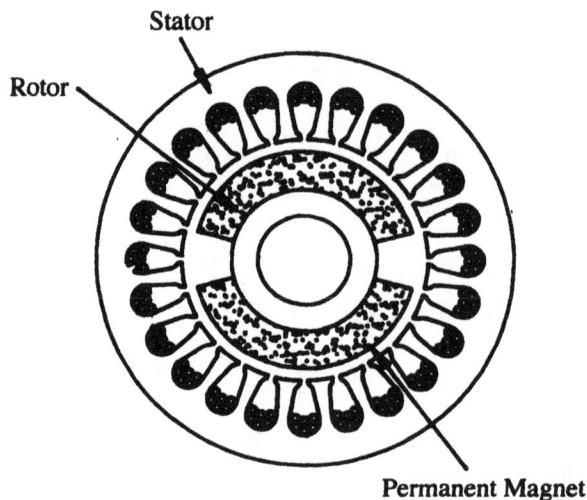

Fig. 7-7

PM synchronous machines are closely related to conventional synchronous machines in performance and configuration, and, the stators of most synchronous machines are identical. Therefore the theoretical analysis of PM synchronous machines as viewed from the stator terminals follows closely conventional synchronous machine theory. The following equations and analysis are on a per-phase basis, as is customary in machine analysis. Subscripts have the following meaning:

d = direct axis component
q = quadrature axis component
m = a mutual reactance component
r = rotor component
l = leakage reactance or resistance component
i = internal component (behind leakage impedance)
o = excitation voltage
c = core loss component

Uppercase symbols are used to represent RMS values. Standard symbols are used for impedances, voltages, currents, and power. In addition, θ = power factor angle; δ = power (or torque) angle; a = number of parallel paths in armature; m = number of phases; Z = total armature conductors; V = terminal voltage; I = terminal current/phase; and K_w = winding factor, which is the product of distribution, pitch, and skew factors.

We consider a synchronous motor neglecting core losses and operating under steady state. Trigonometric considerations of the phasor diagram for an underexcited synchronous motor (Fig. 7-8), result in the following:

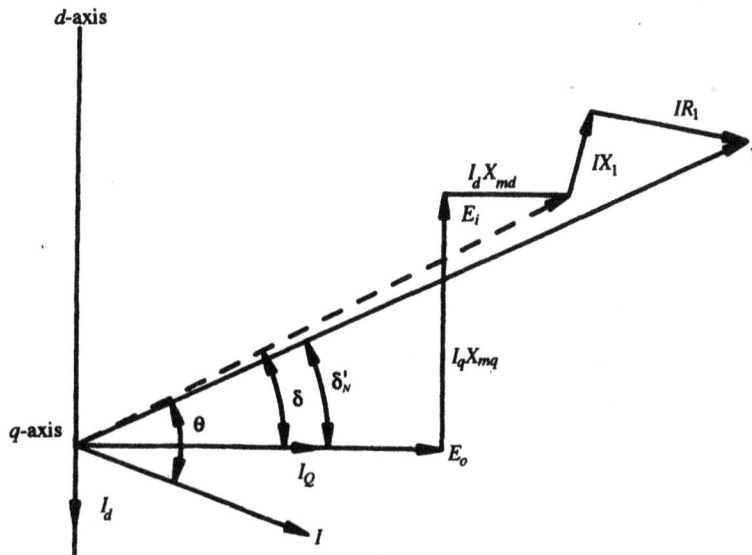

Fig. 7-8

$$V \sin \delta = I_q X_q - I_d R_l$$
$$V \cos \delta = E_o + I_d X_d + I_q R_l \tag{7.13}$$

Similar phasor diagrams may be drawn for synchronous alternators by reversing the sign of the reactance and resistance drops of Fig. 7-8; also, of course, the overexcited case can be considered by assuming (for a motor) a leading power factor. Solving for the component currents gives

$$I_d = \frac{V(X_q \cos \delta - R_l \sin \delta) - E_o X_q}{X_d X_q + R_l^2} \tag{7.14}$$

$$I_q = \frac{V(R_l \cos \delta + X_d \sin \delta) - E_o R_l}{X_d X_q + R_l^2} \tag{7.15}$$

The total input power to m phases is due to the current components in phase with the terminal voltage, V:

$$P_{in} = mV(I_q \cos \delta - I_d \sin \delta) \tag{7.16}$$

which can be modified to eliminate the functions of the torque angle, δ, as

$$P_{in} = m\left[I_q E_o + I_d I_q (X_d - X_q) + I^2 R_l\right] \tag{7.17}$$

The electromagnetic power developed at the airgap can be expressed as

$$P_e = m\left[I_q E_o + I_d I_q (X_{md} - X_{mq})\right] \tag{7.18}$$

which may also be expressed in terms of conventional theory as

$$P_e = m\left[I_q E_{iq} - I_d E_{id}\right] \tag{7.19}$$

Another expression (in terms of the "interior torque angle", δ_i) is given as

$$P_e = m\left[\frac{E_o E_i}{X_{md}} \sin \delta_i + \frac{E_i^2 (X_{md} - X_{mq})}{2 X_{md} X_{mq}} \sin 2\delta_i\right] \tag{7.20}$$

Note that

$$\begin{aligned} X_d &= X_{md} + X_l \\ X_q &= X_{mq} + X_l \end{aligned} \tag{7.21}$$

As a good approximation to (7.20), the terminal—rather than internal—values may be used, which means substituting V for E_i, X_d for X_{md}, δ for δ_i, etc.

There are several interesting points to note in the above analysis. Although this analysis follows closely that of conventional synchronous machines, there are several important differences. In many types of PM machines, including the common interior types, $X_d < X_q$, and in most other types the two reactances are fairly close to each other in magnitude. Both of these conditions are in contrast to the conventional salient-pole synchronous machine. The principal cause of these reactances' relative values is the very low permeability of the PM itself, which is located in the direct axis in interior types and many other configurations. The implications of these reactance values should be noted with reference to (7.20). The second term of this equation is normally known as "reluctance torque," and, in many types of PM machines, this term may be negative. It is only in the transverse types, which are basically reluctance machines, that this term contributes appreciable positive torque. In most other types of PM machines, the first term of (7.20) is the major contributor to torque development. As a corollary to this reactance relationship, the maximum power developed by an interior type—and some other types—of PM machines occurs at a torque angle, δ, greater than 90°, as would be expected, whereas maximum power occurs at torque angles less than 90° in conventional salient-pole machines.

7.5. HYSTERESIS MOTORS

Like the reluctance motor, a hysteresis motor does not have a dc excitation. Unlike the reluctance motor, however, the hysteresis motor does not have a salient rotor. Instead, the rotor of a hysteresis motor has a ring of special magnetic material, such as chrome, steel, or cobalt, mounted on a cylinder of aluminum or some other nonmagnetic material, as shown in Fig. 7-9. The stator of the motor is similar to that of an induction motor, and the hysteresis motor is started as an induction motor.

To understand the operation of the hysteresis motor, we may consider the hysteresis and eddy-current losses in the rotor. We observe that, as in an induction motor, the rotor has a certain equivalent resistance. The power dissipated in this resistance determines the electromagnetic torque developed by the motor, as discussed in Chapter 5. We may conclude that the electromagnetic torque developed by a hysteresis motor has two components—one by virtue of the eddy-current loss and the other because of the hysteresis loss. We know that the eddy-current loss can be expressed as

$$P_e = K_e f_2^2 B^2 \tag{7.22}$$

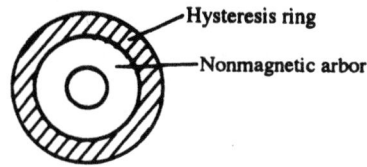

Fig. 7-9

where K_e = a constant
          f_2 = frequency of the eddy currents
          B = flux density

In terms of the slip s, the rotor frequency f_2 is related to the stator frequency f_1 by

$$f_2 = sf_1 \tag{7.23}$$

Thus (7.22) and (7.23) yield

$$p_e = K_e s^2 f_1^2 B^2 \tag{7.24}$$

And the torque T_e is related to p_e by (see Chapter 5)

$$T_e = \frac{p_e}{s\omega_s} \tag{7.25}$$

so that (7.24) and (7.25) give

$$T_e = K's \tag{7.26}$$

where $K' = K_e f_1^2 B^2 / \omega_s$ = a constant.
        Next, for the hysteresis loss, p_h, we have

$$p_h = K_h f_2 B^{1.6} = K_h s f_1 B^{1.6} \tag{7.27}$$

and for the corresponding torque, T_h, we obtain

$$T_h = K'' \tag{7.28}$$

where $K'' = K_h f_1 B^{1.6}/\omega_s$ = a constant.
        Notice that the component, T_e, as given by (7.26), is proportional to the slip and decreases as the rotor picks up speed. It is eventually zero at synchronous speed. This component of the torque aids in the starting of the motor. The second component, T_h, as given by (7.28), remains constant at all rotor speeds and is the only torque when the rotor achieves the synchronous speed.

Solved Problems

7.1.   For a 230-V, 1-phase induction motor, the parameters of the equivalent circuit, Fig. 7-1(a), are $R_1 = R_2' = 8\ \Omega$, $X_1 = X_2' = 12\ \Omega$, and $X_m = 200\ \Omega$. At a slip of 4%, calculate (a) input current, (b) input power, (c) developed power, and (d) developed torque (at rated voltage). The motor speed is 1728 rpm.

        From Fig. 7-1(a):

$$\mathbf{Z}_f = \frac{(j100)\left(\dfrac{4}{0.04} + j6\right)}{j100 + \dfrac{4}{0.04} + j6} = 47 + j50 \ \Omega$$

$$\mathbf{Z}_b = \frac{(j100)\left(\dfrac{4}{1.96} + j6\right)}{j100 + \dfrac{4}{1.96} + j6} = 1.8 + j5.7 \ \Omega$$

$$\mathbf{Z}_1 = R_1 + jX_1 = 8 + j12 \ \Omega$$

$$\mathbf{Z}_{\text{total}} = 56.8 + j67.7 = 88.4 \ \angle 50^\circ \ \Omega$$

(a) $$\text{input current} \equiv I_1 = \frac{230}{88.4} = 2.6 \ \text{A}$$

(b)
$$\text{power factor} = \cos 50^\circ = 0.64 \ \text{lagging}$$
$$\text{input power} = (230)(2.6)(0.64) = 382.7 \ \text{W}$$

(c)   Proceeding as in Problem 5.14(c), we have

$$P_d = [I_1^2 \ Re \ (\mathbf{Z}_f)](1 - s) + [I_1^2 \ Re \ (\mathbf{Z}_b)][1 - (2 - s)]$$

$$= I_1^2 \ [Re \ (\mathbf{Z}_f) - Re \ (\mathbf{Z}_b)](1 - s) = (2.6)^2(47 - 1.8)(1 - 0.04)$$

$$= 293.3 \ \text{W}$$

(d) $$\text{torque} = \frac{P_d}{\omega_m} = \frac{293.3}{2\pi(1728)/60} = 1.62 \ \text{N} \cdot \text{m}$$

7.2.   To reduce the numerical computation, Fig. 7-1(a) is modified by neglecting $0.5 \ X_m$ in $\mathbf{Z}_b$ and taking the backward-circuit rotor resistance at low slips as $0.25R_2'$. With these approximations, repeat the calculations of Problem 7.1 and compare the results.

$$\mathbf{Z}_f = 47 + j50 \ \ \Omega$$

$$\mathbf{Z}_b = 2 + j6 \ \ \Omega$$

$$\mathbf{Z}_1 = 8 + j12 \ \ \Omega$$

$$\mathbf{Z}_{\text{total}} = 57 + j68 = 88.7\angle 50^\circ \ \ \Omega$$

(a) $$I_1 = \frac{230}{88.7} = 2.6 \ \text{A}$$

(b)
$$\cos \phi = 0.64 \ \text{lagging}$$
$$\text{input power} = (230)(2.6)(0.64) = 382.7 \ \text{W}$$

(c) $$P_d = (2.6)^2 \ (47 - 2)(1 - 0.04) = 292.0 \ \text{W}$$

(d)
$$\text{torque} = \frac{292.0}{2\pi(1728)/60} = 1.61 \text{ N} \cdot \text{m}$$

7.3. A 1-phase, 110-V, 60-Hz, 4-pole induction motor has the following constants in the equivalent circuit, Fig. 7-1(a): $R_1 = R_2' = 2 \ \Omega$, $X_1 = X_2' = 2 \ \Omega$, $X_m = 50 \ \Omega$. There is a core loss of 25 W and a friction and windage loss of 10 W. For a 10% slip, calculate (a) the motor input current and (b) the efficiency.

$$\mathbf{Z}_f = \frac{(j25)\left(\dfrac{1}{0.1} + j1\right)}{j25 + \dfrac{1}{0.1} + j1} = 8 + j4 \ \Omega$$

$$\mathbf{Z}_b = \frac{(j25)\left(\dfrac{1}{1.9} + j1\right)}{j25 + \dfrac{1}{1.9} + j1} = 0.48 + j0.96 \ \Omega$$

$$\mathbf{Z}_1 = 2 + j2 \ \Omega$$

$$\mathbf{Z}_{\text{total}} = 10.48 + j6.96 = 12.6\angle 33.6° \ \Omega$$

(a)
$$I_1 = \frac{110}{12.6} = 8.73 \text{ A}$$

(b)   developed power = $(8.73)^2 (8 - 0.48)(1 - 0.10) = 516$ W
output power = $516 - 25 - 10 = 481$ W

input power = $(110)(8.73)(\cos 33.6°) = 800$ W

efficiency = $\dfrac{481}{800} = 60\%$

7.4. What is the relative amplitude of the resultant forward-rotating flux to the resultant backward-rotating flux for the motor of Problem 7.3 operating at 10% slip?

The voltages V_f and V_b indicated in Fig. 7-1(a) are respectively proportional to the amplitudes of the two resultant fluxes. Thus,

$$\frac{\phi_f}{\phi_b} = \frac{V_f}{V_b} = \frac{Z_f}{Z_b} = \frac{\sqrt{(8)^2 + (4)^2}}{\sqrt{(0.48)^2 + (0.96)^2}} = \frac{4}{0.48} = 8.33$$

7.5. For the motor of Problem 7.1, calculate the currents in the various elements of the equivalent circuit when (a) the motor is running on no-load, $s = 0$; and (b) the rotor is blocked, $s = 1$. Both cases are at rated input voltage.

For the two cases, the circuits are shown in Fig. 7-10.

(a) $\mathbf{Z}_f = j100 \ \Omega$

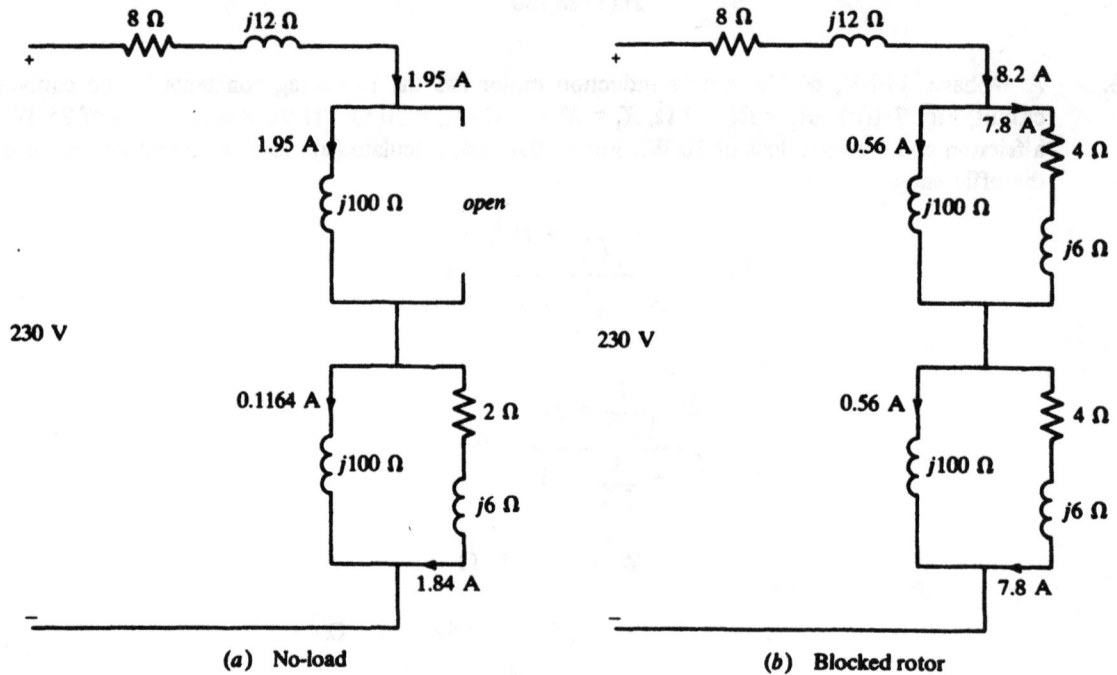

(a) No-load   (b) Blocked rotor

Fig. 7-10

$$Z_b = \frac{(j100)(2 + j6)}{2 + j106} = 1.78 + j5.7 \ \Omega$$

$$Z_1 = 8 + j12 \ \Omega$$

$$Z_{total} = 9.78 + j117.7 = 118\angle 85° \ \Omega$$

$$I_1 = \frac{230\angle 0°}{118\angle 85°} = 1.95\angle -85° \ A$$

$$V_b = I_1 Z_b = (1.95\angle -85°)(5.97\angle 73°) = 11.64\angle -12° \ V$$

$$\text{current in } j100 = \frac{11.64\angle 12°}{100\angle 90°} = 0.1164\angle -102° \ A$$

$$\text{current in } (2 + j6) = \frac{11.64\angle -12°}{6.32\angle 72°} = 1.84\angle -84° \ A$$

These currents are marked in Fig. 7-10(a).

(b)
$$Z_f = Z_b = \frac{(j100)(4 + j6)}{4 + j106} = 3.5 + j5.8 \ \Omega$$

$$Z_1 = 8 + j12 \ \Omega$$

$$Z_{total} = 15 + j23.6 = 28\angle 57° \ \Omega$$

$$I_1 = \frac{230}{28} = 8.2 \ A$$

$$V_f = V_b = I_1 Z_f = (8.2)\sqrt{(3.5)^2 + (5.8)^2} = 56 \text{ V}$$

$$\text{current in } j100 = \frac{56}{100} = 0.56 \text{ A}$$

$$\text{current in } (4 + j6) = \frac{56}{\sqrt{4^2 + 6^2}} = 7.8 \text{ A}$$

These currents are labeled in Fig. 7-10(b).

7.6. Using the computations of Problem 7.5, show how the equivalent circuit parameters of a 1-phase induction motor may be approximately determined from no-load and blocked-rotor tests.

Problem 7.5 shows that in the no-load test $0.5X_m$ (or $j100$) may be neglected in $\mathbf{Z}_b$ because the current in that branch is much smaller than the current in $0.5(R_2' + jX_2')$. Thus, on no-load we have

$$\mathbf{Z}_0 = (R_1 + jX_1) + j0.5X_m + (0.25R_2' + j0.5X_2') \tag{1}$$

Under the blocked-rotor condition, again the magnetizing current is small and X_m may be neglected, giving

$$\mathbf{Z}_s = (R_1 + jX_1) + (R_2' + jX_2') \tag{2}$$

Assuming that $X_1 = X_2'$ and measuring R_1, we can determine X_1, X_2', X_m, and R_2', as demonstrated in Problem 7.7.

7.7. A no-load test on a 1-phase induction motor yielded the following data: input voltage, 110 V; input current, 3.7 A; input power, 49.5 W; friction and windage loss, 7 W. The results of a blocked-rotor test were: input voltage, 48 V; input current, 5.6 A. Assuming $X_1 = X_2'$ and a stator resistance of 2.1 Ω, determine the parameters of the double-revolving-field equivalent circuit, Fig. 7-1(a).

From the no-load test [see (1) of Problem 7.6],

$$Z_0 = 110/3.7 = 29.7 \text{ Ω}$$

$$I_{10}^2 (R_1 + 0.25R_2') = (3.7)^2(2.1 + 0.25R_2') = 49.5 - 7 = 42.5 \text{ W}$$

whence $R_2' = 4.0$ Ω. From the blocked-rotor test, $Z_s = 48/5.6 = 8.56$ Ω, and (2) of Problem 7.6 gives

$$(8.56)^2 = (6.1)^2 + (2X_1)^2$$

from which $X_1 = X_2' = 3$ Ω. Now, using the values of Z_0, R_1, R_2', X_1, and X_2' in (1) of Problem 7.6,

$$(29.7)^2 = (2.1 + 1.0)^2 + (4.5 + 0.5 X_m)^2 \quad \text{or} \quad X_m = 50 \text{ Ω}$$

To summarize: $R_1 = 2.1$ Ω, $R_2' = 4.0$ Ω, $X_1 = X_2' = 3.0$ Ω, and $X_m = 50$ Ω.

7.8. For the magnetic circuit of Fig. 1-20(a) (Chapter 1) we have $l_g = 1$ mm, $l_m = H = 5$ cm, and core cross section = 9 cm^2 (for the entire circuit). The magnet is made of Alnico V (see Fig. 1-11). Calculate the airgap flux density. Assume that the iron portion of the circuit is infinitely permeable, and neglect leakage and fringing.

From Fig. 1-11 (Chapter 1),

$$\mu_r\mu_0 = \frac{1.23 - 1.125}{24 \times 10^3} = 4.375 \times 10^{-6} \ \text{H/m}$$

The recoil line intersects at $(-40 \times 10^3, 0.99)$. Equation of the line is

$$B = 1.165 + 4.375 \times 10^6 \ H \quad -40 \times 10^3 < H < 0$$

The circuit reluctance is:

$$R = R_m + R_g = \frac{5 \times 10^{-2}}{4.375 \times 10^{-6} \times 9 \times 10^{-4}} + \frac{1 \times 10^{-2}}{4\pi \times 10^{-7} \times 9 \times 10^{-4}}$$

$$= 12.7 \times 10^6 + 0.884 \times 10^6 = 13.584 \times 10^6 \ \text{H}^{-1}$$

Thus, the respective fluxes are:

$$\phi_r = B_m A_m = 1.165 \times 9 \times 10^{-4} = 1.049 \times 10^{-3} \ \text{Wb}$$

$$\phi_g = \frac{R_m}{R_m + R_g} \phi_r = \frac{12.7 \times 10^6}{13.584 \times 10^6} \times 1.049 \times 10^{-3} = 0.981 \times 10^{-3} \ \text{Wb}$$

Hence, the airgap flux density is:

$$B_g = \frac{\phi_g}{A_g} = \frac{0.981 \times 10^{-3}}{9 \times 10^{-4}} = 1.09 \ \text{T}$$

7.9. Refer to the magnetic circuit of Fig. 1-20(a) and the data given in Problem 7.8. Obtain a relationship between B_m, the magnet flux density, and H_m, the field intensity in the magnet. Hence define the load line, and graphically determine the airgap flux density.

For the given magnetic circuit we have:

$$H_g l_g + H_m l_m = 0 \qquad H_g = -\frac{l_m}{l_g} H_m$$

$$\phi = B_g A_g = B_m A_m \qquad B_g = \frac{A_m}{A_g} B_m = \mu_0 H_g$$

Thus,

$$B_m = -\mu_0 \left(\frac{A_g}{A_m}\right)\left(\frac{l_m}{l_g}\right) H_m$$

$$= -10\mu_0 H_m = -12.56 \times 10^{-6} \ H_m$$

which is the equation to the load line. From the demagnetization curve $B_g = 1.1$ T.

7.10. In (1.26) (Chapter 1) we have derived an expression for the magnet volume. Now the magnetic circuit of Fig. 1-20 is modified by shaping the "poles" and thus reducing the cross section of the airgap to 6 cm². Determine the minimum magnet volume to obtain a 0.8-T airgap flux density. Use the data given in Problem 7.8.

For minimum magnet volume, the operating point is at the maximum energy product. From Fig. 1-11 (Chapter 1), this operating point is at $B = 1.0$ T and $H_m = -45\,000$ A/m. Now,

$$A_m = \frac{B_g}{B_m} A_g = \frac{0.8}{1.0} \times 6 = 4.8 \text{ cm}^2$$

$$l_m = -l_g \frac{H_g}{H_m} = -1 \times 10^{-3} \frac{B_g}{\mu_0 H_m} = -1 \times 10^{-3} \frac{0.8}{4\pi \times 10^{-7}(-45 \times 10^3)}$$

$$= 1.415 \text{ cm}$$

$$\text{volume} = 1.415 \times 4.8 = 6.8 \text{ cm}^3$$

7.11.   The permanent magnet material used in a motor has a profound effect on the characteristics of the motor. For example, a ceramic (ferrite) magnet results in a motor that operates at a relatively low airgap flux density but that can sustain relatively high levels of armature reaction without demagnetization. Compare the ceramic motor with one using Alnico VI (Fig. 1-11) in the following steps.

    (a)   It is seen that Alnico VI must be operated at a much higher permeance ratio than the ceramic material. Since the permeance ratio is a function of the magnetic permeance of the external magnetic circuit, mainly the airgap length, what does this imply concerning the length of practical airgaps usable in an Alnico VI motor as compared to those in ceramic motors?

    (b)   Assume a permeance ratio (B_d/H_d) of 50 for Alnico VI. What permeance ratio is used in the case of a ceramic magnet?

    (c)   What armature reaction (in terms of a field intensity, H_a) can be tolerated in the Alnico VI motor before the flux density drops to $0.8B_r$? Properties of Alnico VI are $B_r = 10.5$ kG; $H_c = 770$ (Oe); $\mu_{recoil} = 4.9$; and maximum energy product, $G - Oe \times 10^6 = 3.8$.

    (a)   The airgap must, generally, be shorter. A plot of the demagnetization curve for a ceramic magnet is shown in Fig. 7-11, where OA is the operating line. The figure also shows the intrinsic magnetization (defined as the vector difference at a point in the magnet between the magnetic induction at that point and the magnetic induction that would exist in a vacuum under the influence of the same magnetizing force, that is, $\bar{B}_i = \bar{B} - \mu_0 \bar{H}$.

Denoting armature reaction by H_a (Fig. 7-11), OA shifts to MB. At A, we have

$$\text{permeance ratio} = \frac{3280}{700} = 4.7$$

    (b)   The ratio $\dfrac{B_d}{H_d} = 50$; or $H_d = \dfrac{8000}{50} = 160$ Oe; $0.8B_r = 8000$ gauss.

    (c)   Allowable armature reaction = 240 Oe

$$\frac{E_{Alnico}}{E_{ferrite}} = \frac{8000}{3280} = 2.4$$

Electromagnetic power is given by:

$$P_e \propto EI_a = 0.8B_r \times OM \qquad OM \approx \text{armature reaction mmf}$$

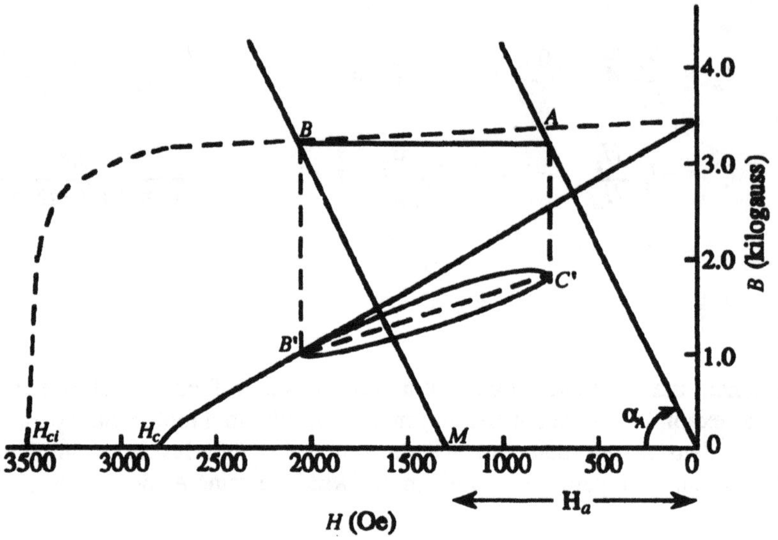

H (Oe)

Fig. 7-11

Thus, for Alnico VI, $P_e \propto 8000 \times 240 = 1.92 \times 10^6$
for ferrite, $P_e \propto 3250 \times 1350 = 4.388 \times 10^6$

And $\dfrac{(P_e)_{\text{ferrite}}}{(P_e)_{\text{Alnico}}} = \dfrac{4.388}{1.92} = 2.285$

7.12. A permanent magnet dc motor develops a maximum torque of 2 N · m at an airgap flux density of 0.6 T. The axial length of the motor is 6 cm and the bore is 10 cm. Calculate the required electrical loading of the armature.

From (*7.11*) we have:

$$2 = \pi \,(10 \times 10^{-2})^2 \; 6 \times 10^{-2} \times 0.6X$$

Solving for X yields $X = 1768$ A/m.

7.13. If the motor of Problem 7.12 has 2 poles and 144 active conductors, calculate the armature current. Also, determine the speed in rpm at which the motor develops a maximum torque, if the armature induced voltage at maximum torque is 36 V.

We have

$$\frac{ZI_a}{2a} = X\pi D \quad \text{or} \quad I_a = \frac{2\pi XaD}{Z}$$

From the data of Problem 7.11 we obtain

$$I_a = \frac{2\pi \times 1768 \times 2 \times 0.1}{144} = 15.43 \text{ A}$$

Now, $T_e \omega_m = EI_a$ gives

$$\omega_m = \frac{2\pi n}{60} = \frac{EI_a}{T_e}, \text{ or}$$

$$n = \frac{60EI_a}{2\pi T_e} = \frac{60 \times 36 \times 15.43}{2\pi \times 2} = 2652 \text{ rpm}$$

7.14. In a 3-phase, 230-V, wye-connected PM synchronous motor, the d-axis and q-axis reactances are equal, each having a per phase value of 1.1 Ω. The armature resistance and the core losses are negligible. Calculate the power developed by the motor if it operates at a 45° power angle, and internal induced voltage/phase is 127 V.

From (7.20) (with the suggested approximations) we obtain:

$$P_d = 3\left[\frac{V E_o}{X_d} \sin \delta + \frac{V_t^2(X_d - X_q)}{2X_d X_q} \sin 2\delta\right]$$

$$= 3 \times \frac{230}{\sqrt{3}} \times \frac{127}{1.1} \sin 45° + 0 = 32.5 \text{ kW}$$

7.15. Repeat Problem 7.14 if the armature resistance is 0.2 Ω/phase and the internal induced voltage/phase is 100 V.

From (7.15), with $V = 230/\sqrt{3} = 132.8$ V, we obtain

$$I_q = \frac{132.8\,(0.2 \cos 45° + 1.1\ 45°) - 100 \times 0.2}{1.1 \times 1.1 + 0.2^2} = 81.66 \text{ A}$$

Because $x_d = x_q$, the second term in (7.18) is zero. Thus,

$$P_d = 3I_q E_o = 3 \times 81.66 \times 100 = 24.5 \text{ kW}$$

Supplementary Problems

7.16. A single-phase induction motor has an air-gap field, produced by the main winding, which is given by

$$B(\theta, t) = B_m \cos 2\theta \sin 377t$$

What is its synchronous speed, in rpm?   *Ans.* 1800 rpm

7.17. The equivalent circuit parameters of a 230-V, 1-phase, 6-pole, 60-Hz induction motor are: $R_1 = R_2' = 10$ Ω, $X_1 = X_2' = 10$ Ω, and $X_m = 100$ Ω. At a slip of 5%, calculate (a) motor speed, (b) input current, (c) power factor, (d) developed torque.
Ans. (a) 1140 rpm; (b) 3.68 A; (c) 0.5 lagging; (d) 1.84 N · m

7.18. The input power to the motor of Problem 7.17 at 230 V, while running on no-load, is 31.4 W at 1.2 A. What is the approximate efficiency of the motor at 5% slip?   *Ans.* 44.2%

7.19. Repeat the calculations of Problem 7.17 by neglecting $0.5X_m$ in Z_b [Fig. 7-1(a)] and taking the backward-circuit rotor resistance to be $0.25R_2'$.   *Ans.* (a) 1140 rpm; (b) 3.65 A; (c) 0.5 lagging; (d) 1.77 N · m

7.20. A 110-V, 1-phase, 4-pole, 60-Hz induction motor has the following circuit constants: $R_1 = R_2' = 1.6\ \Omega$, $X_1 = X_2' = 1.8\ \Omega$, $X_m = 60\ \Omega$; core loss is 16 W and friction and windage loss is 12 W. Determine the efficiency of the motor at 0.04 slip by using (a) the exact equivalent circuit and (b) the approximate equivalent circuit (as described in Problem 7.19).   *Ans.* (a) 0.69; (b) 0.68

7.21. The stator resistance of a 1-phase induction motor is 2.5 Ω and its leakage reactance is 2.0 Ω. On no-load, the motor takes 4 A at 96 V and at 0.25 lagging power factor. The no-load friction and windage loss is negligible. Under the blocked-rotor condition, the input power is 130 W at 6 A and 42 V. Obtain the equivalent circuit parameters.   *Ans.* $R_1 = 2.5\ \Omega$; $R_2' = 3.5\ \Omega$; $X_1 = 2\ \Omega$; $X_2' = 1.6\ \Omega$; $X_m = 50\ \Omega$

7.22. The reluctance motor of Fig. 3-17(a) is excited by a voltage source

$$v = V_m \cos \omega t$$

Show that the time-average torque developed by the motor is given by

$$T_{\text{average}} = \frac{V_m^2}{4\omega} \left(\frac{1}{X_q} - \frac{1}{X_d} \right) \sin 2\delta$$

where X_d and X_q are, respectively, the d- and q-axis reactances, and δ is the power angle.

7.23. A permanent magnet dc motor draws negligible current on no-load, while running at 3300 rpm at 110 V. The armature circuit resistance is 1.1 Ω. Calculate the motor speed at 55 V, if the electromagnetically developed torque is 1.0 N · m.   *Ans.* 1546 rpm.

7.24. A permanent magnet disk-type dc generator is shown in Fig. 7-12. The disk has an outer radius r_1 and an inner radius r_2. If it rotates in a uniform magnetic field B at a speed ω_m, derive an expression for the voltage induced in the disk.   *Ans.* $\frac{1}{2} B\omega_m(r_1^2 - r_2^2)$

7.25. If the disk of Fig. 7-12 is fed with a current I, what is the torque developed by the machine?   *Ans.* $\frac{1}{2} BI(r_1^2 - r_2^2)$

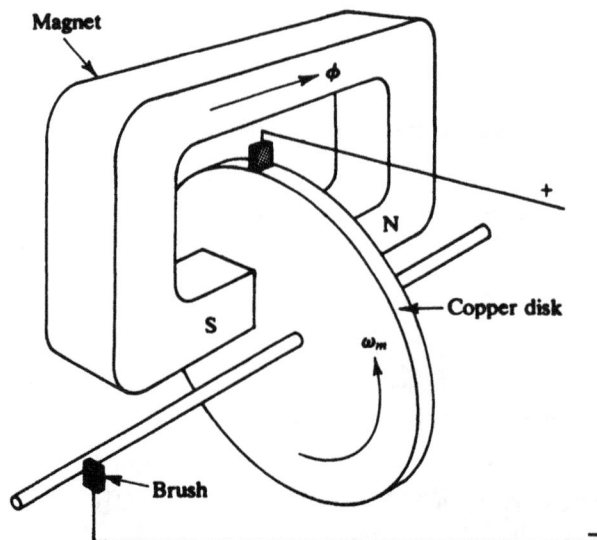

Fig. 7-12

7.26. The disk of Fig. 7-12 has the following dimensions: $r_1 = 80$ mm, $r_2 = 10$ mm, $b = 5$ mm. The conductivity of the disk material is 5.7 MS/m. If the disk rotates at 6000 rpm while delivering a current of 3150 A in a uniformly distributed field flux of 45 mWb, calculate (a) the induced voltage, (b) the electromagnetic torque, (c) the terminal voltage, (d) the electrical power loss, (e) the output power.
Ans.   (a) 4.5 V;   (b) 22.56 N · m;   (c) 4.463 V;   (d) 115 W;   (e) 14 kW.

7.27. A PM 3-phase Y-connected synchronous motor may be considered as a round-rotor machine, having a synchronous reactance of 0.9 Ω/phase. When run as a generator, the no-load terminal voltage is 60 V. Calculate the power angle if the machine develops 3-kW power while running as a motor and supplied at 65 V.   *Ans.* 43.8°

7.28. What is the maximum power that is developed by the motor of Problem 7.27?   *Ans.* 4.33 kW

Chapter 8

Electronic Control of Motors

8.1 GENERAL CONSIDERATIONS

The purpose of a motor control system is to govern one or more of the following parameters: shaft speed, shaft angular position, shaft acceleration, shaft torque, and mechanical output power. Since it is the output mechanical parameters of the motor that are being controlled by input electrical parameters, the peculiar characteristic of the individual machine—that is, the particular relationship it supports between input electrical quantities and output mechanical quantities—is of vital importance in the design and analysis of electronic control. Figure 8-1 is the basic scheme of electronic motor control. This figure illustrates a total motor system including load and power source. The feedback loops are shown by dashed lines, since many motor control schemes are "open-loop." The primary concern of this chapter will be with the box labeled "Controller."

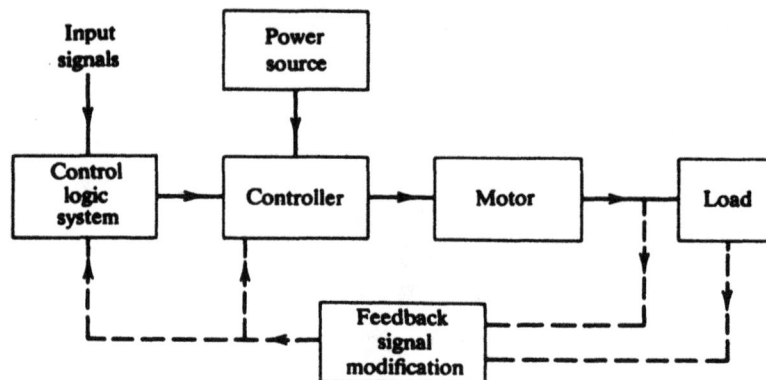

Fig. 8-1

Most methods of motor control involve switching operations—switches may be required to be opened or closed to achieve the desired goal. Modulation of power by turning switches on or off can be accomplished by mechanical switches, such as contactors, or by solid-state electronic switches, such as transistors and thyristors. Because power levels in electric motors and power systems are high compared to those in conventional electronic circuits (such as amplifiers, oscillators, etc.), the study of electronic circuits pertinent to electric machines and power systems is known as power electronics. Thus the scope of power electronics includes the applications of solid-state switches to the control and modulation of power in electric motors and electric power systems.

There is a great variety of solid-state components and systems used to control electric motors. In terms of analysis and applications, no other aspect of electric machines has undergone such dramatic changes in recent years or holds greater potential for improving machine characteristics in the future than does the solid-state control of electric machines.

In this chapter we discuss the various solid-state devices used in power electronics. This discussion will be from a circuit viewpoint. Next, we review waveform analysis because invariably the output waveforms from solid-state switching devices are nonsinusoidal. This is followed by several dc and ac motor control schemes, including illustrative problems on thyristor commutation (or turn-off) techniques.

8.2 POWER SOLID-STATE DEVICES

Many types of solid-state devices exist which are suitable for power electronics applications. In most cases, the maximum ratings such as voltage, current, and time of response of the device are also given. However, these capabilities are seldom achievable simultaneously in a single device. In practice, a device is chosen primarily either for its voltage or current rating. The devices listed here are *pn*-junction devices, having two layers as in a silicon rectifier; three layers, as in a power transistor; or four layers, as in a thyristor. Let us now consider these devices in some detail. [*Note*: In the symbolic representation of the device, we have A = anode; G = gate; K = cathode; B = base; C = collector; E = emitter.]

Silicon Rectifier

Silicon rectifiers are high-power diodes capable of operating at high junction temperatures. The principal parameters of a silicon rectifier are the repetitive peak reverse voltage (PRV) or blocking voltage, average forward current, and maximum operating junction temperature ($\approx 125°C$). The terminal *iv* characteristic of a typical silicon rectifier is shown in Fig. 8-2, which also shows the switching characteristic of an ideal diode. The silicon rectifier has a forward voltage drop of about 1 V at all current levels within its rating.

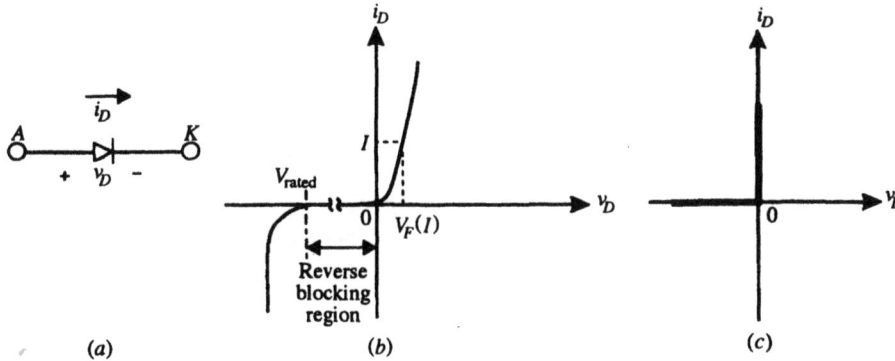

Fig. 8-2

Two of the major applications of silicon rectifiers in power electronics are as freewheeling diodes (providing a bypass for the flow of current) in motor controllers and, in general, as rectifiers.

Silicon-Controlled Rectifiers or Thyristors

A silicon-controlled rectifier (or *SCR*), also known as a *thyristor*, is a four-layer *p-n-p-n* semiconductor switch. Unlike the diode, which has only two terminals—anode and cathode—the thyristor has three terminals—anode, cathode, and gate. The reverse characteristic of the thyristor is similar to that of silicon rectifier just discussed. However, the forward conduction of a thyristor can be controlled by utilizing the gate. Normally, a thyristor will not conduct in the forward direction unless it is "turned on" by applying a triggering signal to the gate. However, the full conduction in a thyristor is not instantaneous. We define the turn-on time, t_{on}, when the anode current reaches 90 percent of its final value. Once the thyristor starts to conduct, it continues to do so until turned off by external means. The turn-off of the thyristor is known as *commutation*. A thyristor is symbolically represented in Fig. 8-3(*a*) and has the *iv* characteristics as shown in Fig. 8-3(*b*), (*c*). The maximum rating of the thyristor is in the range of 5000 V and 3000 A.

Triacs

The triac, often called a bidirectional switch, is approximately equivalent to a pair of back-to-back or antiparallel thyristors fabricated on a single chip of semiconductor material. Triggered conduction may occur in both directions, that is, the triac is a quasibilateral device. Triac applications include light dimming,

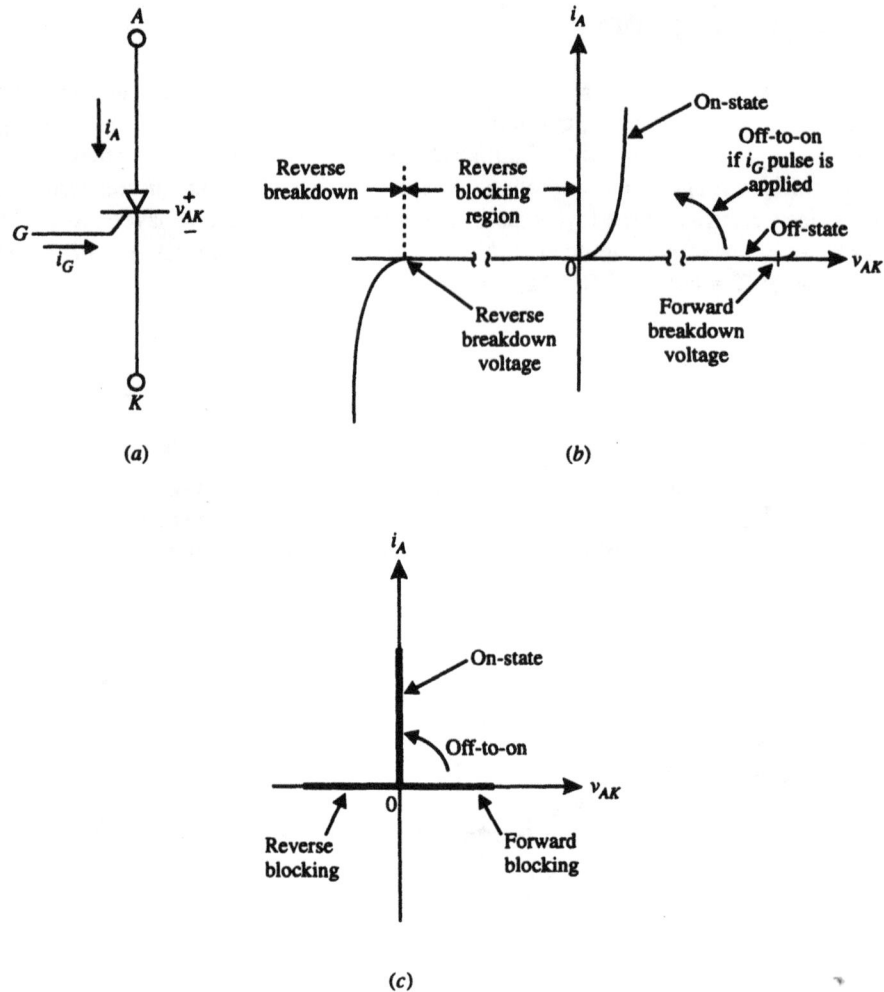

(a)   (b)

(c)

Fig. 8-3

heater control, and ac motor speed control. It should be noted that the triac is a three-terminal device with only one gate, which has an effect on its time response compared with that of two distinct thyristors connected in antiparallel position. The turn-off time of a triac is in the same order of magnitude as that of a thyristor. This implies that a time period approximately equal to the turn-off time must be observed before applying reverse voltage to a triac. In an antiparallel pair, however, reverse voltage can be immediately applied after cessation of forward current in one thyristor. Triacs are not available in as high voltage and current ratings as thyristors at the present time and therefore are used in control of motors of relatively low power ratings.

Symbolically, a triac is shown in Fig. 8-4 and has a rating in the range of 1000 V and 2000 A, with a response time of 1 μs.

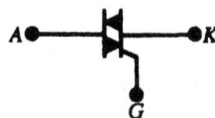

Fig. 8-4

Diverse Thyristors

In addition to the triac, other forms of thyristors include the following:

1. *Gate turn-off thyristor* (GTO). This thyristor can be turned off at a high temperature, and normal commutation circuit is therefore not required. This type of a thyristor has a high blocking voltage rating and is capable of handling large currents.

2. *Gate-assisted thyristor* (GAT). This thyristor requires large power for triggering. It has a small turn-off time and is specially suited for series-type inverters.

3. *Light-activated thyristor* (LAT or LASCR). This thyristor is turned on by photon or light. Such thyristors find application in high-voltage dc transmission.

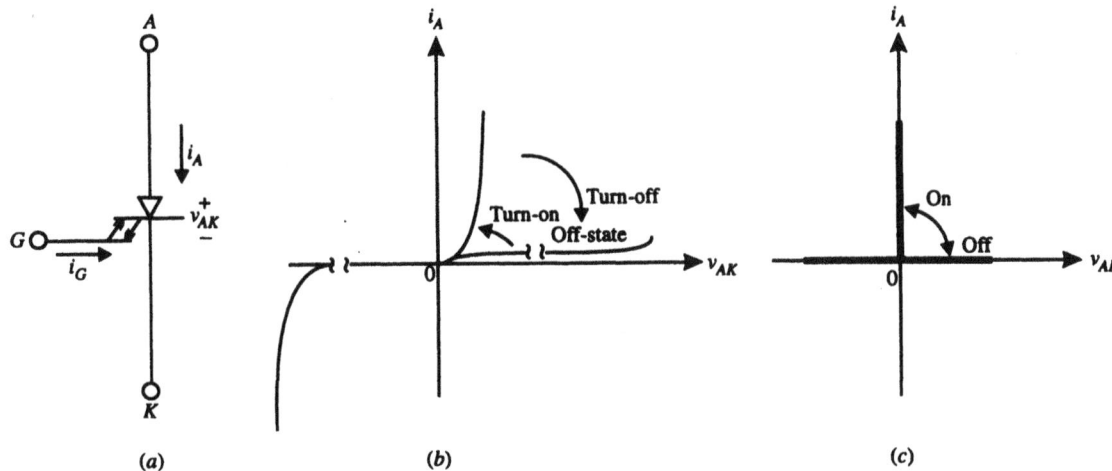

Fig. 8-5

Symbolic representation of these types of thyristors is shown in Fig. 8-5. These thyristors have voltage ratings in the range of 400 to 1000 V with a current rating of about 200 A and a response time of 0.2 to 2.0/μs.

Thyristors, when turned on, have the *i-v* characteristics similar to that shown in Fig. 8-5(*b*) and (*c*).

Power Transistors

When used in motor control circuits, power transistors are almost always operated in a switching mode. The transistor is driven into saturation and the linear gain characteristics are not used. The common-emitter configuration is the most common, because of the high power gain in this connection. The collector-emitter saturation voltage, $V_{CE(SAT)}$ for typical power transistors is from 0.2 to 0.8 V. This range is considerably lower than the on-state anode-to-cathode voltage drop of a thyristor. Therefore, the average power loss in a power transistor is lower than that in a thyristor of equivalent power rating. The switching times of power transistors are also generally faster than those of thyristors, and the problems associated with turning off or commutating a thyristor are almost nonexistent in transistors. However, a power transistor is more expensive than a thyristor of equivalent power capability. In addition, the voltage and current ratings of available power transistors are much lower than those of existing thyristors. It has already been stated that the maximum ratings of a power semiconductor are generally unobtainable concurrently in a single device. This is particularly true of power transistors. Devices with voltage ratings of 1000 V or above have limited current ratings of 10 A or less. Similarly, the devices with higher current ratings, 50 A and above, have voltage ratings of 200 V or less. For handling motor control requiring large current ratings at 200 V or below, it has been common to parallel transistors of lower current rating. This requires great care to assure equal sharing of collector currents and proper synchronization of base currents among the paralleled devices. Some of the commonly used power transistors are as follows.

Bipolar Junction Transistor (BJT)

The symbol and the *i-v* characteristics of a bipolar junction transistor (BJT) are shown in Fig. 8-6. The BJT requires a sufficiently large base current such that

$$I > \frac{I_c}{h_{FE}} \qquad\qquad (8.1)$$

where h_{FE} (≈ 5 to 10) is the dc gain of the BJT to turn it fully on, and I_c is the collector current. In this case the voltage V_{CE} is about 1 to 2 V. A BJT is a current-controlled device and requires a continuous base current to operate in the on-state. A BJT may have a rating of about 1400 V and 200 A.

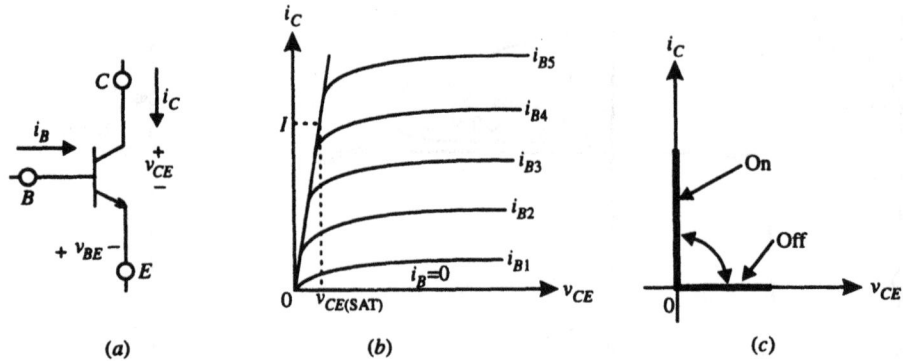

Fig. 8-6

Power Darlington

This designation generally refers to the well-known Darlington-connected transistor pair fabricated on a single chip. The same characteristics are, of course, achievable through the use of two discrete transistors, albeit usually in a larger, more complex, and more costly package. The principal merit of the Darlington device is its high current gain. The operating parameters and failure modes discussed earlier for transistors are also applicable to the Darlington.

Darlington amplifiers are used both in choppers for dc commutator motor control and in inverters for ac motor control, generally for lower-power applications. Recently, larger devices have been developed with ratings as high as 200 A and 100 V or 100 A and 450 V and have been applied to the control of traction motors used in lift trucks and industrial electric vehicles. Current gains as high as 1600 A have been achieved at these high current levels.

Two Darlington configurations are shown in Fig. 8-7.

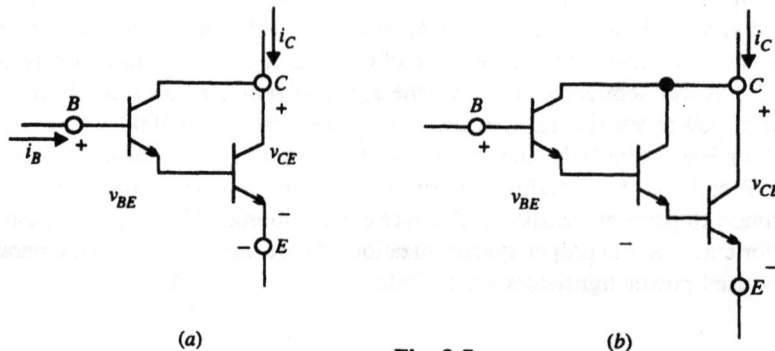

Fig. 8-7

Metal-Oxide-Semiconductor Field Effect Transistor (MOSFET)

The symbol of a MOSFET and its characteristics are shown in Fig. 8-8. With a sufficiently large and continuous gate-source voltage, the MOSFET turns on. MOSFETs have ratings of over 1000 V and 100 A. However, the large current rating comes at similar voltages.

Insulated Gate Bipolar Transistor (IGBT)

The symbol for an IGBT and its i-v characteristics are shown in Fig. 8-9. IGBTs have some of the advantages of the MOSFET, the BJT, and the GTO combined. However, MOSFETs have higher switching speeds. The ratings of IGBTs are up to 1200 V and 100 A.

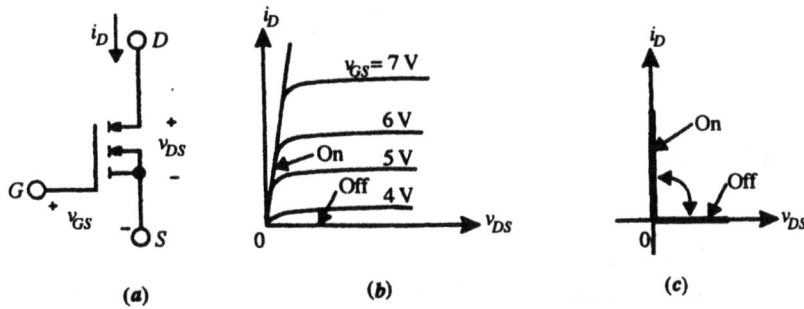

Fig. 8-8

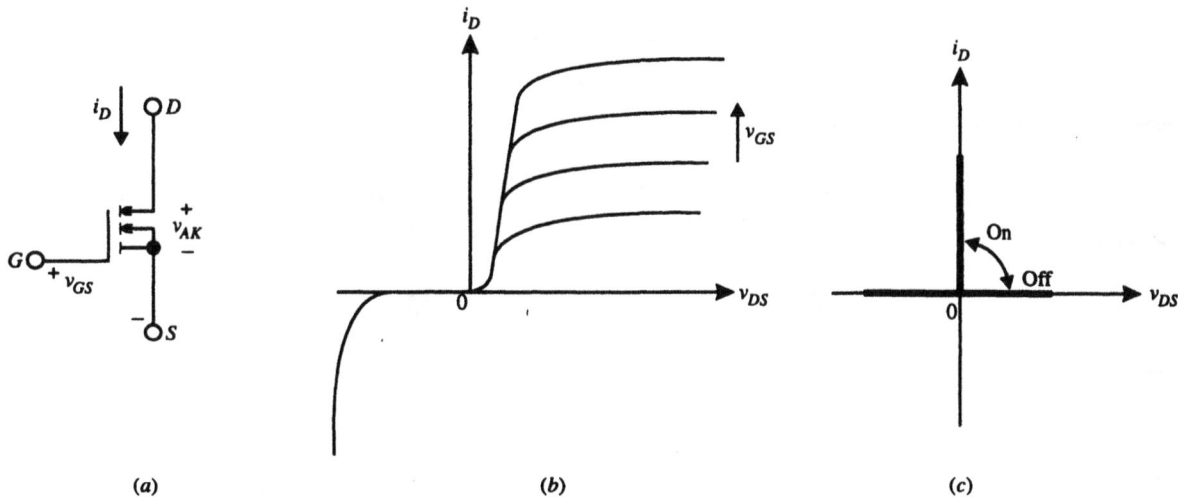

Fig. 8-9

8.3 RMS AND AVERAGE VALUES OF WAVEFORMS

A characteristic of electronic control systems in motor controls, power system protection, and high-voltage dc transmission is that pertinent voltage and current waveforms are nonsinusoidal and often discontinuous. Furthermore, these waveforms change as a function of the level of operation.

The calculation of average and rms values of voltage and current is quite important in electronic control systems for calculating motor power and torque, for heating of wires and other components, and for sizing components and instrumentation. To make these calculations, it is often necessary to return to the definitions of average and rms values, which are respectively defined as

$$A_{ave} = \frac{1}{T_0} \int_0^{T_0} a\, dt \qquad (8.2)$$

$$A_{rms} \equiv A = \left(\frac{1}{T_0} \int_0^{T_0} a^2\, dt \right)^{1/2} \qquad (8.3)$$

where a represents instantaneous value of the parameter and T_0 is the period over which the average (or rms) value is evaluated. In motor control circuits involving power semiconductors, T_0 is usually the "on-time" duration. The fundamental frequency of the signal referred to above is defined by

$$f_p = \frac{1}{T_p} \qquad (8.4)$$

where T_p is the length of a full period.

8.4 CONTROL OF DC MOTORS

Control of dc motors is accomplished by using SCRs to modulate the input voltages to the armature and/or the field circuit of the motor. For an ac source, phase-controlled rectifiers are employed; for a dc source, choppers. But before we discuss these, it is worthwhile to consider the analysis of some passive RL-circuits involving diodes or SCRs.

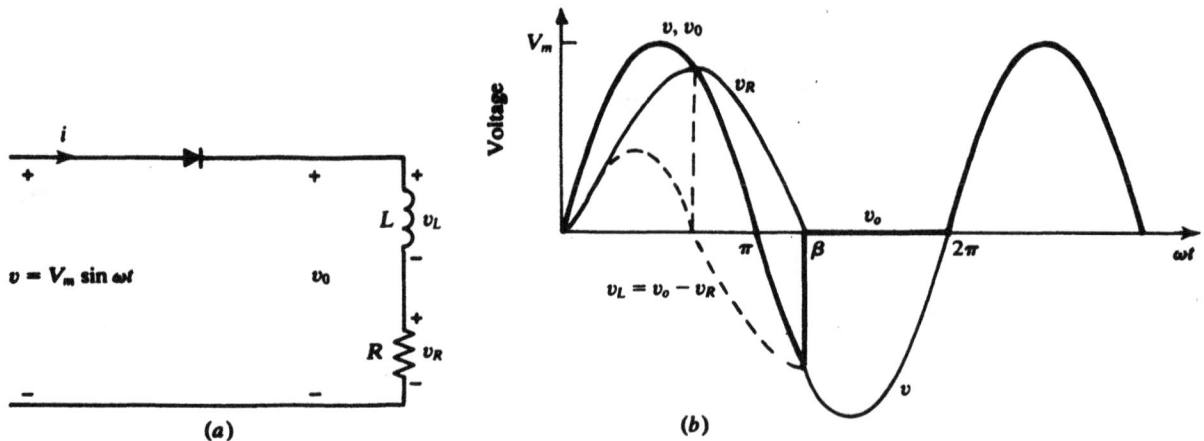

Fig. 8-10

1. *Half-wave rectifier with RL-load.* A half-wave rectifier with *RL*-load is indicated in Fig. 8-10(a). It may be shown (see Problem 8.23) that the current during one period of the applied voltage $v = V_m \sin \omega t$ is given by

$$i = \begin{cases} \dfrac{V_m}{Z} \left[\sin(\omega t - \phi) + e^{-(R/L)t} \sin \phi \right] & 0 < \omega t < \beta \\[2mm] 0 & \beta < \omega t < 2\pi \end{cases} \qquad (8.5)$$

where

$$Z = \sqrt{R^2 + (\omega L)^2} \qquad \tan \phi = \frac{\omega L}{R}$$

The *extinction time* (the time at which the diode stopped conducting) is β/ω; and β can be found from the condition that i be continuous at the extinction time. Thus:

$$\sin(\beta - \phi) + e^{-\beta \cot \phi} \sin \phi + 0 \qquad\qquad (8.6)$$

a transcendental equation for β. The average value of $i(t)$ over a period $2\pi/\omega$ is (see Problem 8.2)

$$I_{avg} = \frac{V_m}{2\pi R} (1 - \cos \beta) \qquad\qquad (8.7)$$

Because the average voltage across the inductor is zero, the average voltage across the load is given by

$$V_{o\,avg} = V_{R\,avg} = RI_{avg} = \frac{V_m}{2\pi} (1 - \cos \beta) \qquad\qquad (8.8)$$

The four voltage waveforms are shown in Fig. 8-10(b)

2. *Half-wave rectifier with dc-motor load.* The circuit is shown in Fig. 8-11(a), where R and L are respectively the armature-circuit resistance and inductance, and e' is the motor back emf, assumed constant. The circuit analysis leads to the following expression for the current:

$$i = \begin{cases} 0 & 0 < \omega t < \alpha \\ \dfrac{V_m}{Z}\left[\sin(\omega t - \phi) + Be^{-(R/L)t}\right] - \dfrac{e'}{R} & \alpha < \omega t < \beta \\ 0 & \beta < \omega t < 2\pi \end{cases} \qquad (8.9)$$

where

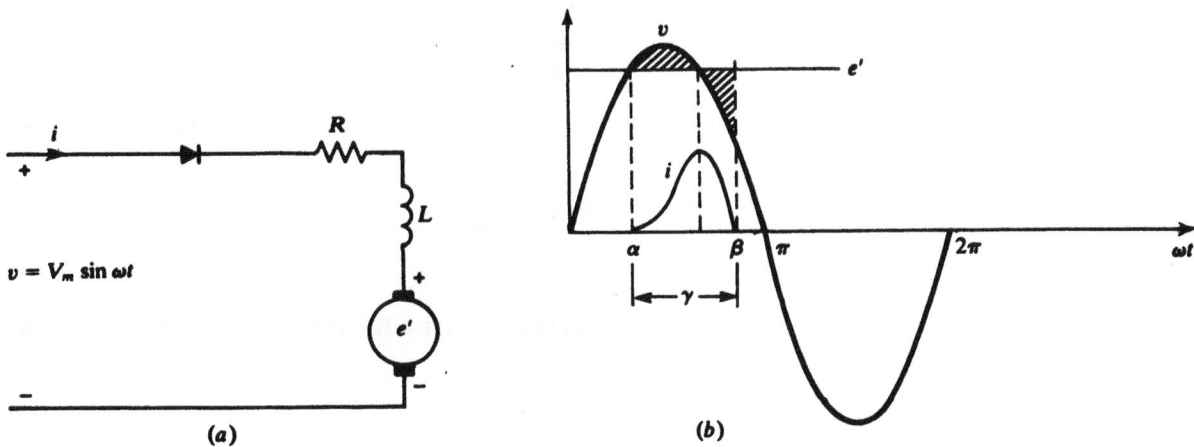

Fig. 8-11

$$Z = \sqrt{R^2 + (\omega L)^2} \qquad \tan \phi = \frac{\omega L}{R}$$

and where

$$B \equiv \left[\frac{e'}{V_m \cos \phi} - \sin(\alpha - \phi) \right] e^{\alpha R/\omega L} \qquad (8.10)$$

is such as to make i continuous at $\omega t = \alpha$. It is seen from (8.9) that the diode starts conducting at $\omega t = \alpha$; the *firing angle*, α, is determined by the condition $v = e' + 0$, i.e.,

$$\sin \alpha = \frac{e'}{V_m} \qquad (8.11)$$

As is shown in Fig. 8-11(b), conduction does not necessarily stop when v becomes less than e'; rather, it ends at $\omega t = \beta$, when the energy stored in the inductor during the current buildup has been completely recovered. The *extinction angle*, β, may be determined from the continuity of (8.9) at $\omega t = \beta$; we find

$$\sin(\beta - \phi) + Be^{-\beta \cot \phi} = \frac{\sin \alpha}{\cos \phi} \qquad (8.12)$$

as the transcendental equation for β, in which B is known from (8.10). The average value of the current over one period of the applied voltage is found to be

$$I_{avg} = \frac{1}{R} V_{Ravg} = \frac{V_m}{2\pi R} (\cos \alpha - \cos \beta - \gamma \sin \alpha) \qquad (8.13)$$

where $\gamma \equiv \beta - \alpha$ is the *conduction angle*. Figure 8-10(b) shows the waveforms.

SCR-Controlled DC Motor

In the example above, the dc-motor load was not controlled by the half-wave rectifier; the back emf remained constant, implying that the motor speed was unaffected by the cyclic firing and extinction of the diode. To achieve a control, we use a thyristor instead of the diode, as shown in Fig. 8-12(a). The corresponding waveforms are illustrated in Fig. 8-12(b). The motor torque (or speed) may be varied by varying α. Explicitly, for the armature we integrate

$$v_m = Ri + L \frac{di}{dt} + e \qquad (8.14)$$

over the conduction period $\alpha/\omega < t < \beta/\omega$, during which v_m coincides with the line voltage v. The result is:

$$V'_m = \frac{V_m(\cos \alpha - \cos \beta)}{\gamma} = RI' + E' \qquad (8.15)$$

where a prime indicates an average over the conduction period. Over a full period of the line voltage, the average armature current is given by

$$I_{avg} = \frac{\gamma}{2\pi} I' \qquad (8.16)$$

and the average torque is given by

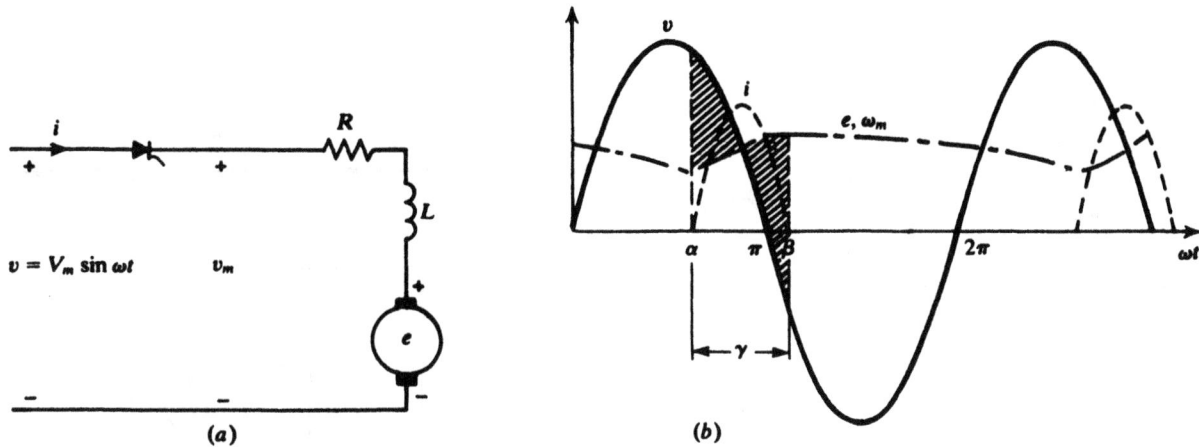

Fig. 8-12

$$T_{avg} = kI_{avg} = \frac{k\gamma}{2\pi} I' \tag{8.17}$$

Equations (8.15), (8.16), and (8.17) govern the steady-state performance of a thyristor-controlled dc motor.

Chopper-Controlled DC Motor

A motor-chopper simplified circuit, and the corresponding voltage and current waveforms are given in Fig. 8-13. Observe that when the thyristor turns off, the applied voltage, v_m drops from V_t to zero. However, armature current i continues to flow through the path completed by the freewheeling diode until all the energy stored in L has been dissipated in R. Then v_m becomes equal to the motor back emf and stays at that value until the thyristor is turned on, whereupon it regains the value V_t.

If the speed pulsations are small, then the motor back emf may be approximated by its average value, $k\Omega_m$, yielding

$$L\frac{di}{dt} + Ri + k\Omega_m = v_m = \begin{cases} V_t & 0 < t < \alpha\lambda \\ 0 & \alpha\lambda < t < \gamma\lambda \\ k\Omega_m & \gamma\lambda < t < \lambda \end{cases} \tag{8.18}$$

as the electrical equation of the system. Here, λ is the period of the thyristor signal, α is the fraction of the period over which the thyristor is conductive (the *duty cycle*), and γ is the fraction of the period over which armature current flows.

Equation (8.18) has the solution, subject to the initial condition $i(0) = 0$,

$$i = \begin{cases} \frac{V_t - k\Omega_m}{R}(1 - e^{-t/\tau}) & 0 < t < \alpha\lambda \\ \frac{k\Omega_m}{R}[e^{(\gamma\lambda - t)/\tau} - 1] & \alpha\lambda < t < \gamma\lambda \\ 0 & \gamma\lambda < t < \lambda \end{cases} \tag{8.19}$$

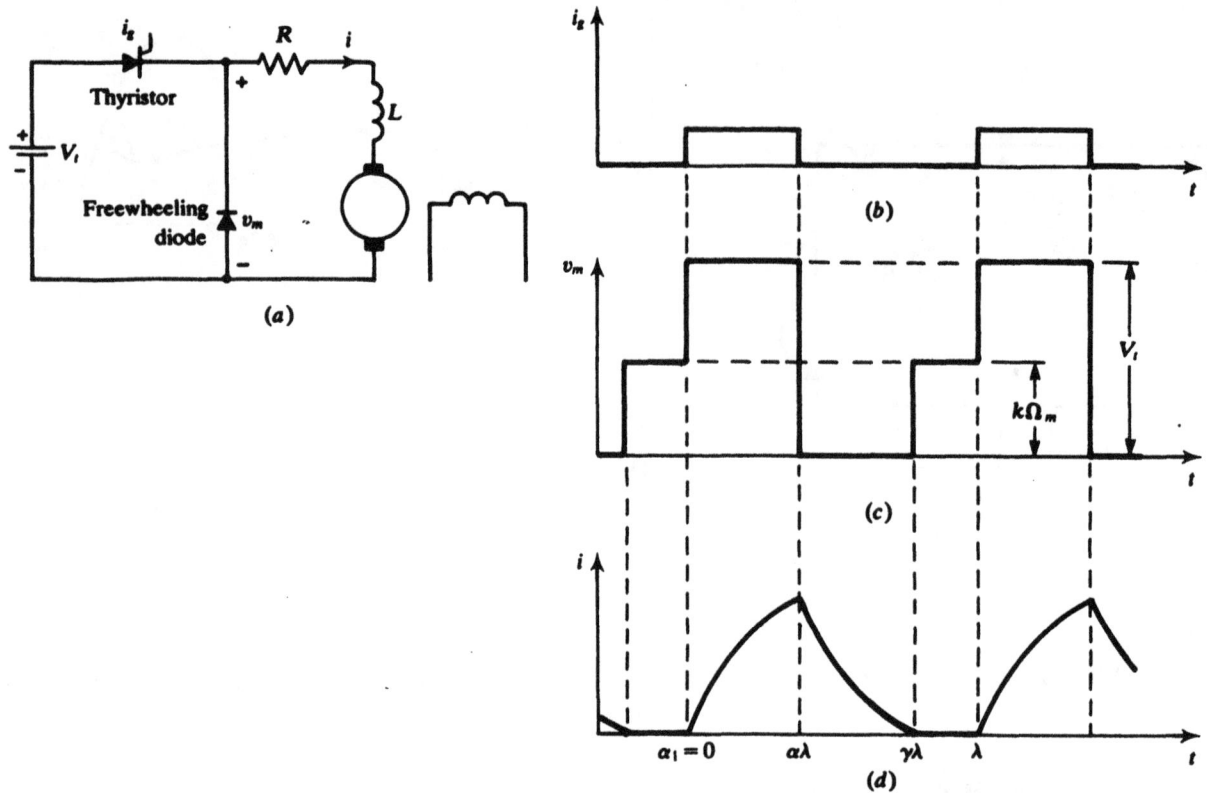

Fig. 8-13

where $\tau \equiv L/R$ is the armature time constant. Continuity of (8.19) at $t = \alpha\lambda$ yields the equation for λ:

$$\gamma = \frac{\tau}{\lambda} \ln\left(1 + \frac{e^{\alpha\lambda/\tau} - 1}{\Omega^*}\right) \qquad (8.20)$$

in which $\Omega^* \equiv k\Omega_m/V_t$ is the normalized (dimensionless) average rotational speed of the motor. Now, it is apparent that when α is sufficiently large and Ω^* sufficiently small, (8.20) gives $\gamma > 1$, which is impossible. Thus, we must distinguish between two modes of operation of the machine.

Mode I is defined by all (α, Ω^*)-combinations satisfying

$$1 > \frac{\tau}{\lambda} \ln\left(1 + \frac{e^{\alpha\lambda/\tau} - 1}{\Omega^*}\right)$$

In this mode, γ is given by (8.20) and the armature current, (8.19), vanishes over a fraction $1 - \gamma$ of the basic cycle.

Mode II is defined by all (α, Ω^*)-combinations satisfying

$$1 \leq \frac{\tau}{\lambda} \ln\left(1 + \frac{e^{\alpha\lambda/\tau} - 1}{\Omega^*}\right)$$

If the equality holds, (8.19) is valid with $\gamma = 1$; that is, the armature current becomes zero only at the period points. If the inequality holds, (8.19) is no longer valid. The governing differential equation, (8.18), and the boundary conditions, must be changed to admit a strictly positive solution, for which again $\gamma = 1$.

The average torque average speed characteristic of the motor can now be derived. Integration of (8.18) over one period of the thyristor signal gives

$$RI_{avg} + k\Omega_m = \alpha V_t + k\Omega_m(1 - \gamma)$$

On the other hand, the torque equation of the motor,

$$J\dot{\omega}_m + b\omega_m + t_0 = ki$$

where t_0 is the load torque, b is the rotational friction coefficient, and J is the moment of inertia, integrates to give

$$b\Omega_m + T_{0avg} = kI_{avg} \qquad (8.22)$$

Eliminating I_{avg} between (8.21) and (8.22), we obtain the desired relation between $T^* \equiv T_{0avg}/(kV_t/R)$, the normalized (dimensionless) average torque, and $\Omega^* \equiv k\Omega_m/V_t$:

$$T^* = \alpha - \left(\frac{bR}{k^2} + \gamma\right)\Omega^*$$

that is,

$$T^* = \begin{cases} \alpha - \left[\frac{bR}{k^2} + \frac{\tau}{\lambda}\ln\left(1 + \frac{e^{\alpha\lambda/\tau} - 1}{\Omega^*}\right)\right]\Omega^* & \text{in Mode I} \\ \alpha - \left(\frac{bR}{k^2} + 1\right)\Omega^* & \text{in Mode II} \end{cases} \qquad (8.23)$$

Figure 8-14 indicates the appearance of the torque-speed curves for several values of α. Observe the linearity of the curves in the region corresponding to Mode II, which is separated from the Mode I region by the dashed curve.

8.5 CONTROL OF AC MOTORS

The speed control of most ac motors is accomplished through inverters or cycloconverters. Small, single-phase ac motors are sometimes controlled by TRIACs, which regulate the phase of the input voltage to the motor.

Inverters

An inverter converts dc to ac at a desired voltage and frequency. The output voltage waveform is nonsinusoidal and the harmonics tend to have an adverse overall effect on the motor performance. There are, however, methods available for the reduction of the harmonics.

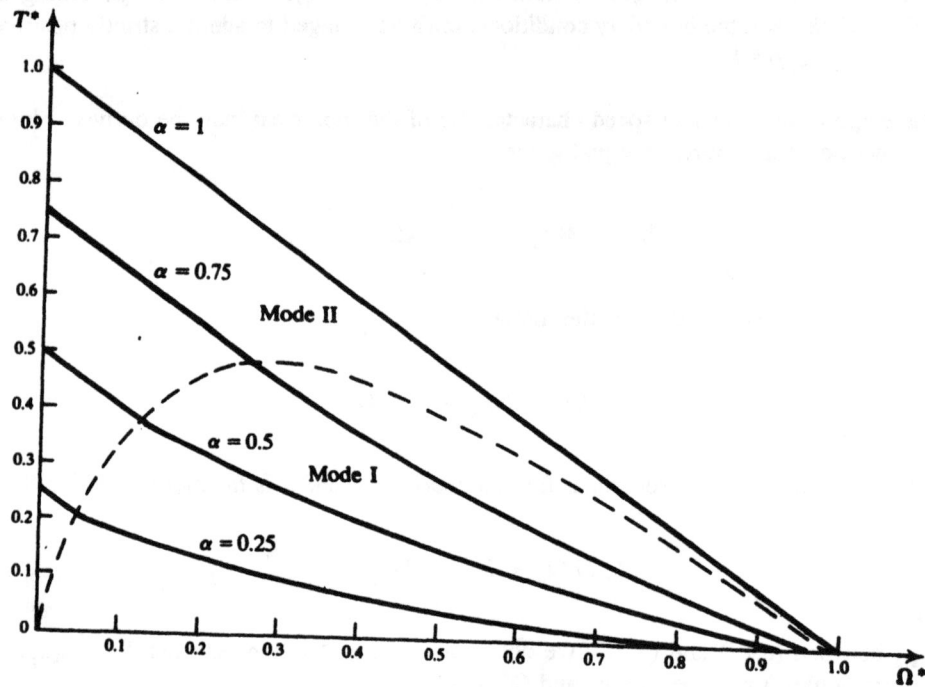

Fig. 8-14

A simplified single-phase, half-bridge inverter is shown in Fig. 8-15(a). Clearly, the output frequency ω will depend on the gating period T, via $\omega T = 2\pi$. The circuit equation is

$$L \frac{di}{dt} + Ri = v = \begin{cases} V/2 & 0 < t < T/2 \\ -V/2 & T/2 < t < T \end{cases} \qquad (8.24)$$

Expressing v in a Fourier series

$$v = \frac{2V}{\pi} \sum_{n \text{ odd}} \frac{1}{n} \sin n\omega t \qquad (8.25)$$

we solve (8.24) by superposition, to obtain the steady-state current:

$$i = \frac{2V}{\pi} \sum_{n \text{ odd}} \frac{1}{nZ_n} \sin (n\omega t - \phi_n) \qquad (8.26)$$

where

$$Z_n = \sqrt{R^2 + (n\omega L)^2} \qquad \tan \phi_n = \frac{n\omega L}{R}$$

At the instants of commutation ($t = 0, T/2, T, 3T/2, 2T, \ldots$),

$$i = \mp I_m \equiv \mp \frac{2V}{\pi\omega L} \sum_{n \text{ odd}} \frac{1}{n^2 + (R/\omega L)^2} = \mp \frac{V}{2R} \tanh \frac{R\pi}{2\omega L} \qquad (8.27)$$

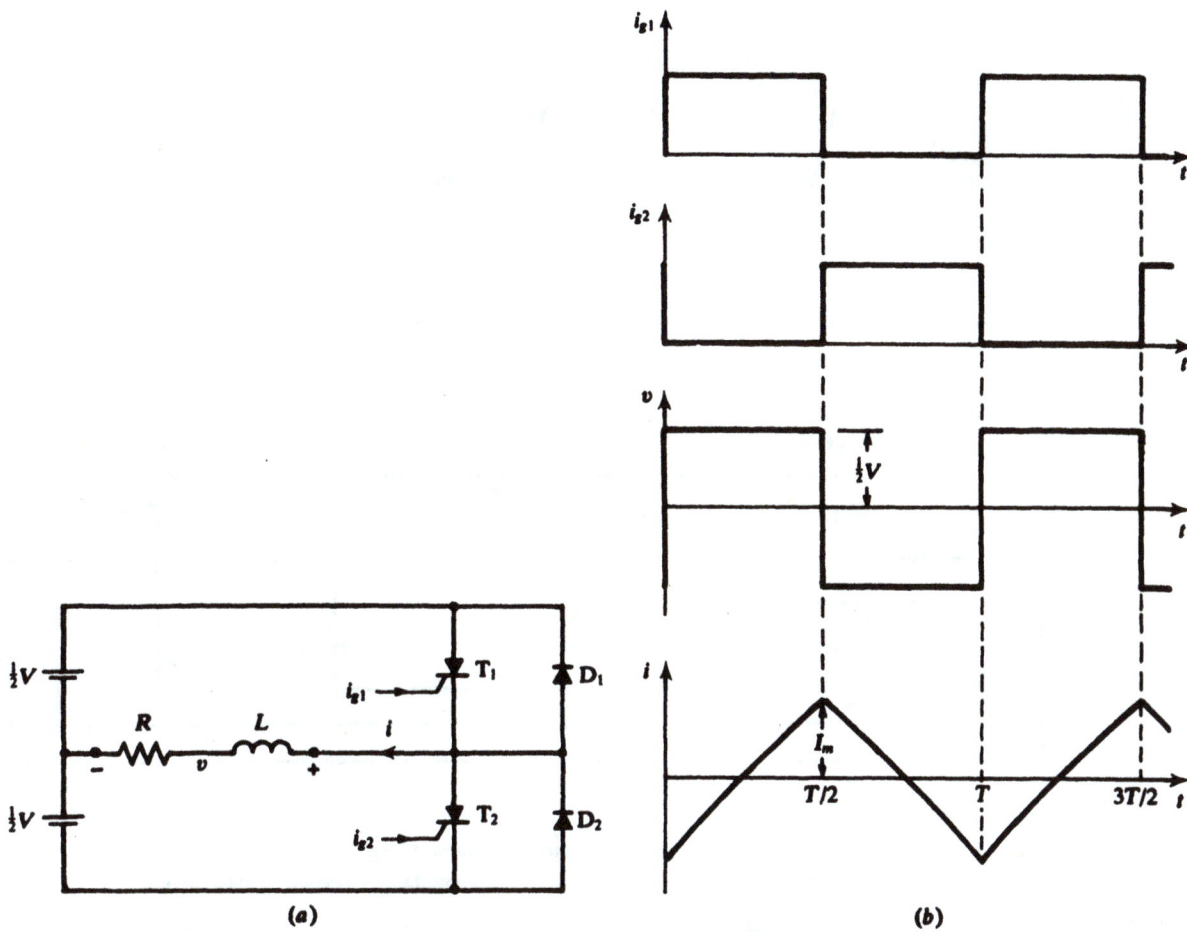

Fig. 8-15

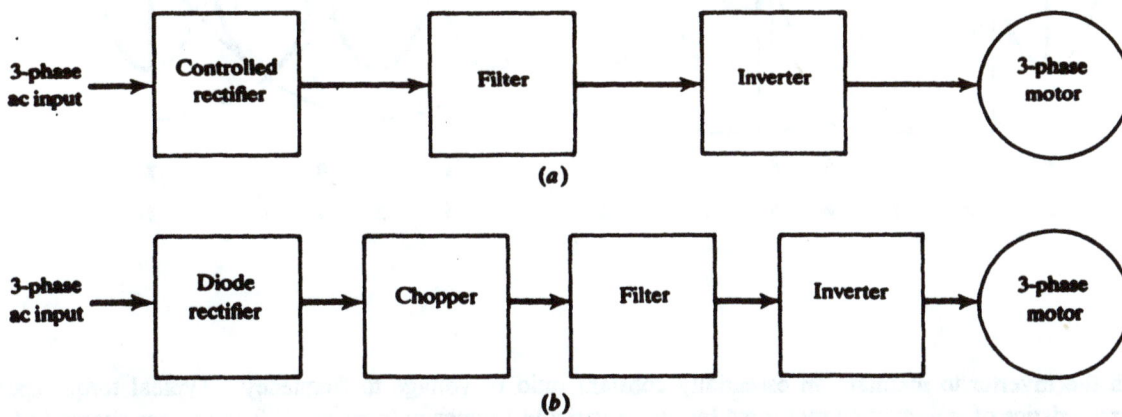

Fig. 8-16

When the plus sign holds in (8.27), forced commutation (Section 8.6) is required; it may not be necessary when the minus sign holds.

   Functional block diagrams of two schemes of speed control of an ac motor by frequency variation are given in Fig. 8-16, where a mechanism such as a controlled rectifier or a chopper is used in conjunction

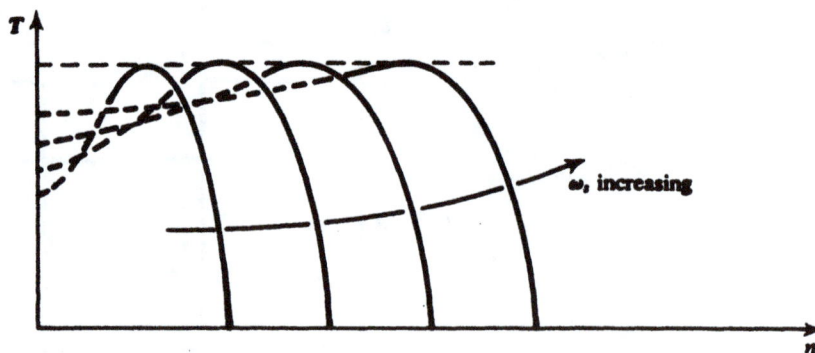

Fig. 8-17

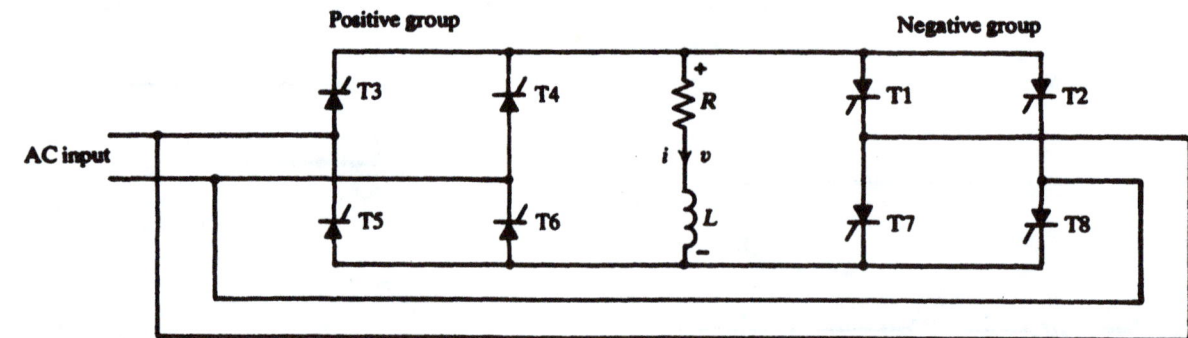

(a)

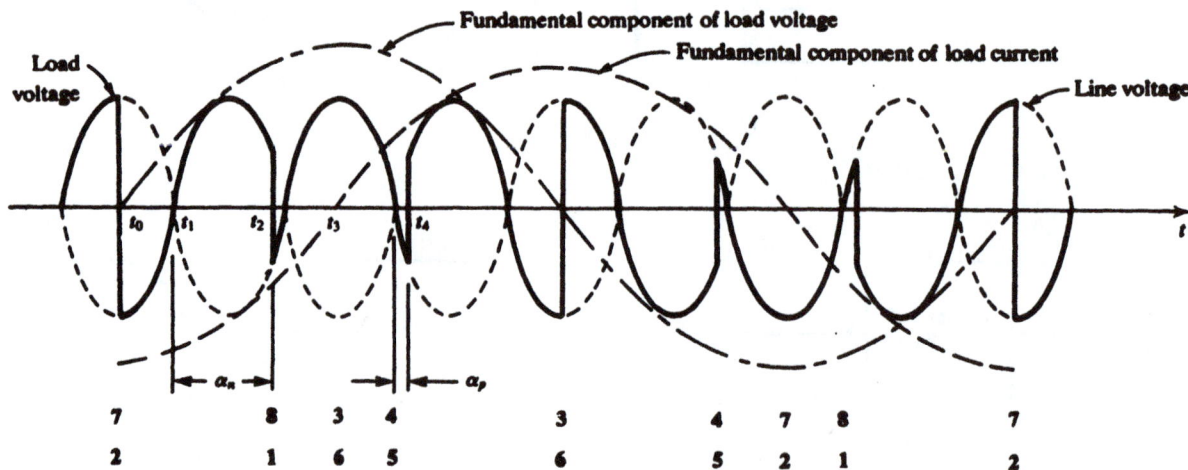

(b)

Fig. 8-18

with the inverter to maintain an essentially constant ratio of voltage to frequency. Typical torque-speed characteristics of an induction motor fed by such a variable-frequency (constant V/f) source are shown in Fig. 8-17.

Cycloconverters

The cycloconverter is a control device used on variable-speed motors supplied by an ac power source. It is a means of converting a source at fixed (peak) voltage and fixed frequency to an output with variable voltage and variable frequency. The source frequency must be at least three to four times the maximum frequency of the output. A single-phase, bridge cycloconverter is shown in Fig. 8-18(a), and the various waveforms are shown in Fig. 8-18(b), which also indicates the firing sequence of the thyristors. In Fig. 8-18(b), α_p denotes the minimum delay time for the positive group of converters, and α_n denotes the maximum delay time for the negative group. The variation of the delay controls the output voltage, in that it determines how many half-cycles of line voltage go to make up one half-cycle of the load voltage fundamental.

8.6 SCR COMMUTATION

As mentioned in Section 8.2 commutation of an SCR refers to the process of turning it off; once turned on, an SCR cannot offer resistance to the forward current unless that current is reduced to zero and held at zero for a period at least equal to the turnoff time. The three basic methods of commutation are as follows.

Line Commutation

In this case, the source is ac and in series with the SCR. The current goes through zero in a cycle and if it remains zero for a period greater than the turnoff time, the SCR will be turned off until a gate current is applied again in the positive voltage cycle. (See Problem 8.14).

Load Commutation

Owing to the nature of the load, the current in the SCR may go to zero and thereby turn it off. This type of commutation is useful mainly in dc circuit. (See Problems 8.15 and 8.16).

Forced Commutation

Forced commutation is achieved in systems energized from dc sources by an arrangement of energy storage elements (capacitors and inductors) and by additional switching devices (usually SCRs). In systems energized from ac sources, forced commutation is brought about by means of the cyclic potential reversal of the power source. The mechanism of forced commutation will be explained with the aid of Figs. 8-19 and 8-20. Figure 8-19 shows the voltage-current relationship that must exist in an SCR during commutation; this characteristic applies to T_1 of Fig. 8-20, with i_1 the anode current in T_1 and v_1 the anode-cathode potential difference.

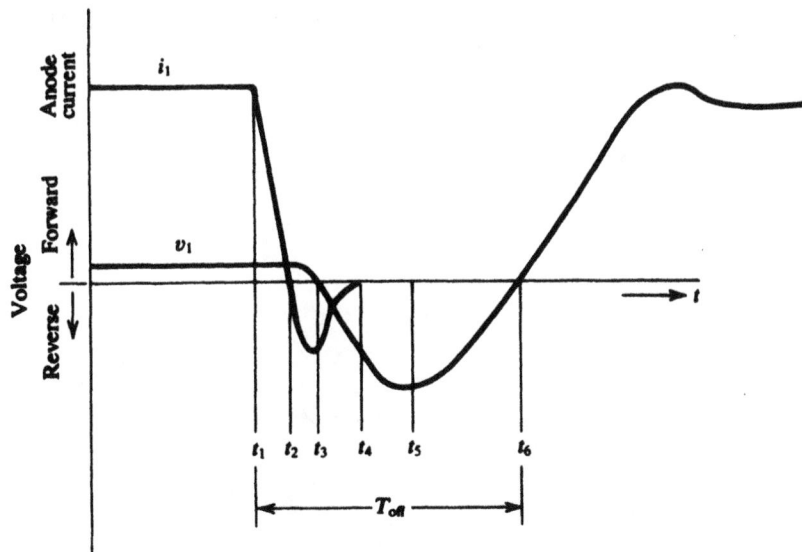

Fig. 8-19

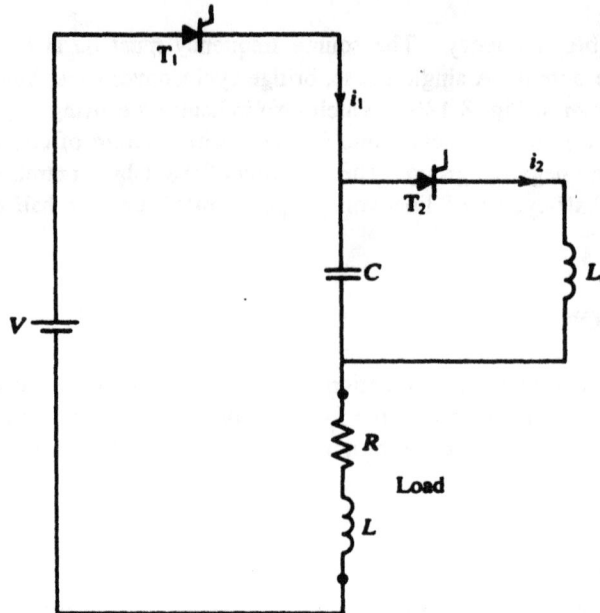

Fig. 8-20

The SCR is in a certain on-state for $t < t_1$, at which moment commutation is initiated by the introduction of a negative voltage into the external anode-cathode circuit. The anode-cathode voltage drop remains at the low on-state level (1.5 to 2.0 V) until the anode current decreases to zero at time t_2; then the voltage begins to go negative. The anode must be maintained at a negative (reversed-biased) potential for a period $T_{off} = t_6 - t_1$, called the *circuit time-off period*, that is somewhat larger than the SCR turnoff time. Following zero anode current, there is a reverse recovery interval, $t_4 - t_2$, of the order of 3 μs.

Solved Problems

8.1.   Find the rms and average values of the voltage waveform shown in Fig. 8-21.

We have

$$a(t) = \begin{cases} \dfrac{1}{T_o}(\zeta - 1)A_{mt} + A_m & 0 < t < T_o \\ 0 & T_o < t < T_p \end{cases}$$

and so

$$A_{avg} = \frac{1}{T_p}\int_0^{T_o}\left[\frac{1}{T_o}(\zeta - 1)A_m t + A_m\right]dt = A_m K_A \frac{T_o}{T_p}$$

with $K_A = (1 + \zeta)/2$, and

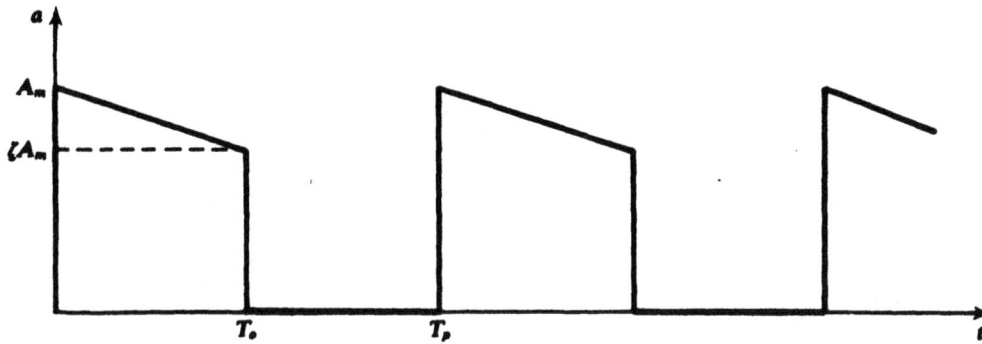

Fig. 8-21

$$A = \left\{ \frac{1}{T_p} \int_0^{T_p} \left[\frac{1}{T_o^2} (\zeta - 1)^2 A_m^2 t^2 + \frac{2}{T_o} (\zeta - 1) A_m^2 + A_m^2 \right] dt \right\}^{1/2} = A_m \sqrt{K_f \frac{T_o}{T_p}}$$

with $K_f = (1 + \zeta + \zeta^2)/3$.

8.2. Verify that the average current in a half-wave rectifier with RL-load is as given by (8.7).

From Fig. 8-5(a)

$$L \frac{di}{dt} + Ri = v_o$$

Averaging over one period $T = 2\pi/\omega$ of the line voltage,

$$\frac{L}{T} \int_0^T \frac{di}{dt} dt + RI_{avg} = \frac{1}{T} \int_0^T v_o dt$$

The integral on the left equals $i(T) - i(0) = 0$, since i is also periodic, of period T; the integral on the right equals

$$\int_0^{\beta/\omega} V_m \sin \omega t \, dt + \int_{\beta/\omega}^T 0 \, dt = \frac{V_m}{\omega} (1 - \cos \beta)$$

Hence

$$RI_{avg} = \frac{V_m}{2\pi} (1 - \cos \beta)$$

which is equivalent to (8.7) and (8.8).

8.3. An RL-load of $2 + j2 \, \Omega$ is connected in series with a diode across an ac source of 110 V. Calculate the average voltage across the resistor.

To apply (8.8) we must know β. For a given ϕ, β can be obtained from (8.6). The solution to (8.6) is graphically expressed in Fig. 8-22, from which $\beta = 225°$ for $\phi = 45°$, which is the phase angle of the load. Then, from (8.8)

$$V_{Ravg} = \frac{\sqrt{2} \, (110)}{2\pi} (1 - \cos 225°) = 42.3 \text{ V}$$

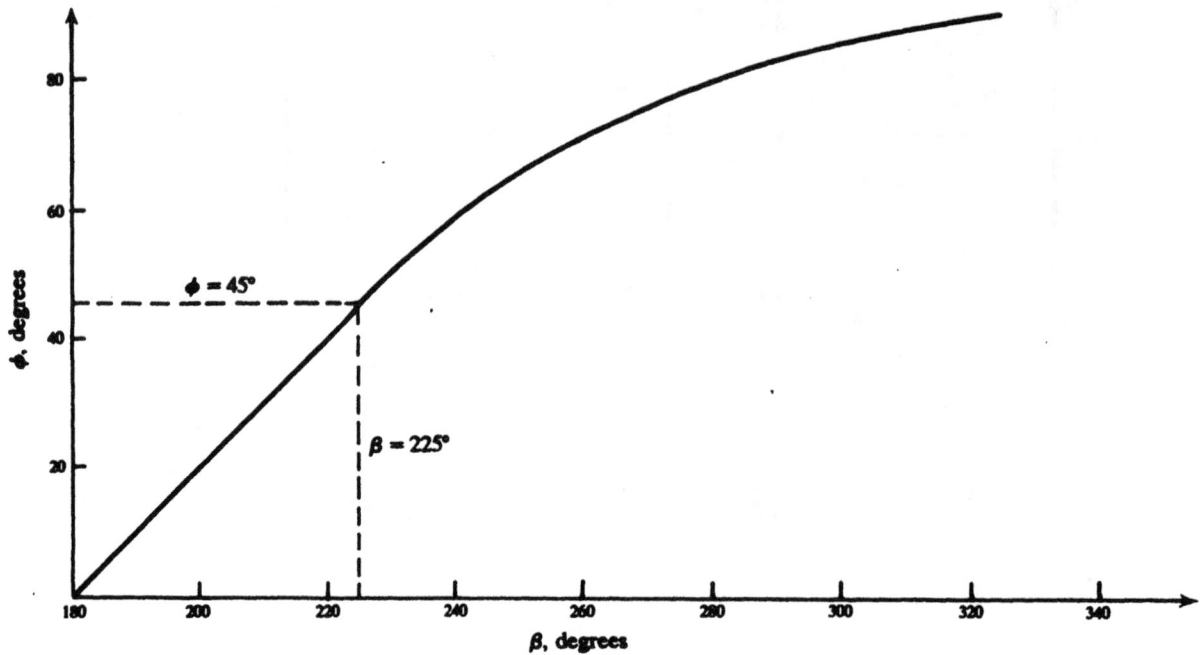

Fig. 8-22

8.4.   A dc motor having an armature resistance of 0.51 Ω and an inductance of 0.78 mH is connected in series with a diode to a 110-V, 60-Hz ac source. The motor is running at an essentially constant speed of 970 rpm. The motor back-emf constant has the value 0.08 V/rpm. Determine the average value of the armature current.

The desired current is given by (8.13), with α determined from (8.11) and β from (8.12). A graphical solution of (8.12) is presented in Fig. 8-23, where $m = \sin \alpha$ and $\gamma = \beta - \alpha$. From the data,

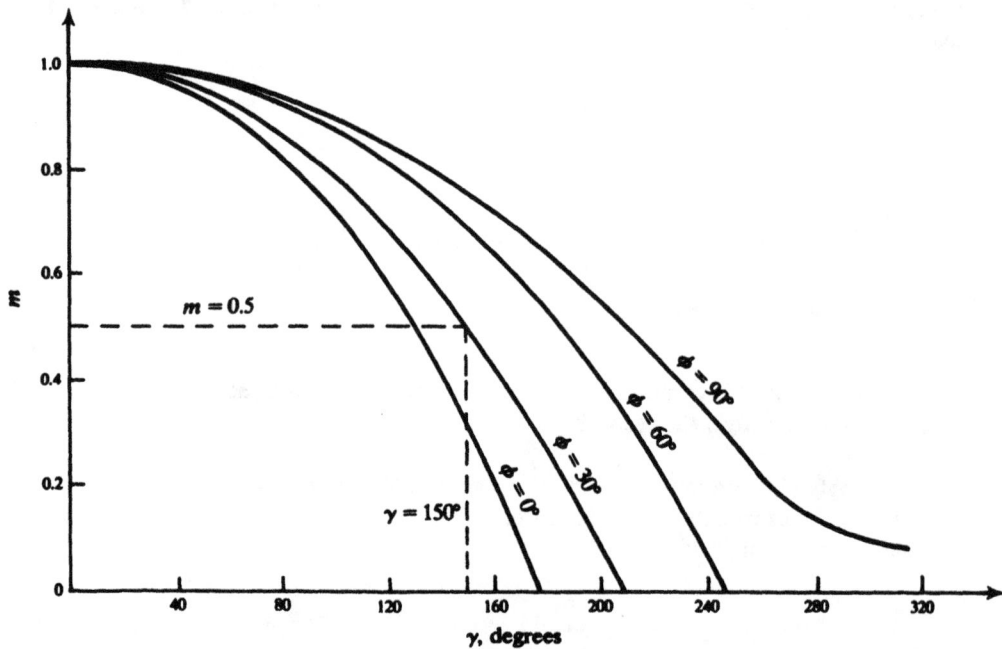

Fig. 8-23

$$\tan \phi = \frac{\omega L}{R} = \frac{(120\pi)(0.78 \times 10^{-3})}{0.51} = 0.577 \quad \text{or} \quad \phi = 30°$$

$$\sin \alpha = \frac{e'}{V_m} = \frac{(0.08)(970)}{110\sqrt{2}} = 0.5 \quad \text{or} \quad \alpha = 30°$$

Hence, from Fig. 8-23, $\gamma = 150°$ and $\beta = \alpha + \gamma = 180°$. Then:

$$I_{avg} = \frac{110\sqrt{2}}{2\pi(0.51)} \left[\cos 30° - \cos 180° - \left(\frac{150}{180}\pi\right)(0.5) \right] = 27 \text{ A}$$

8.5. Find the average torque developed by the motor of Problem 8.4.

$$T_{avg} = kI_{avg} = \left(\frac{0.08}{2\pi/60}\right)(27) = 20.6 \text{ N} \cdot \text{m}$$

(Note the conversion of k to the proper units.)

8.6. If the inductance is neglected in Problem 8.4, determine (a) the firing angle, (b) the conduction angle, (c) the average and rms values of the armature current, (d) the power delivered by the ac source.

(a) $\alpha = 30°$

(b)   If L is negligible, conduction must cease as soon as the line voltage v becomes smaller than the back emf e'. Thus [see Fig. 8-11(b)],

$$180° - \beta = \alpha \quad \text{or} \quad \gamma = \beta - \alpha = 180° - 2\alpha = 120°$$

where γ is the conduction angle. This result may also be read from Fig. 8-23 ($\phi = 0°$).

(c)   From Fig. 8-11(a), or by a limiting process in (8.9), we have for $L \to 0$:

$$i = \begin{cases} 0 & 0 < \omega t < \alpha \\ \dfrac{V_m \sin \omega t - e'}{R} & \alpha < \omega t < \beta \\ 0 & \beta < \omega t < 2\pi \end{cases}$$

where, from (a) and (b), $\alpha = \pi/6$ and $\beta = 5\pi/6$. From (8.13),

$$I_{avg} = \frac{110\sqrt{2}}{2\pi(0.51)} \left[\cos(\pi/6) - \cos(5\pi/6) - (2\pi/3)\sin(\pi/6)\right] = 33.25 \text{ A}$$

and, by integration, using $e' = V_m \sin \alpha = V_m/2$,

$$I^2 = \frac{1}{2\pi} \int_\alpha^\beta \left(\frac{V_m \sin \theta - e'}{R}\right)^2 d\theta = \frac{V_m^2}{2\pi R^2} \int_{\pi/6}^{5\pi/6} \left(\sin \theta - \frac{1}{2}\right)^2 d\theta$$

$$= \frac{V_m^2}{2\pi R^2}\left(\frac{2\pi - 3\sqrt{3}}{4}\right) = \frac{V_m^2}{R^2}(0.0433)$$

Hence

$$I = \frac{110\sqrt{2}}{0.51}(0.0433)^{1/2} + 63.44 \text{ A}$$

(d)   The power input from the source is the sum of the power lost in the resistor and the power taken by the motor:

$$P_{avg} = I^2 R + e'I_{avg} = (63.44)^2(0.51) + \left(\frac{110\sqrt{2}}{2}\right)(33.25) = 4.63 \text{ kW}$$

8.7.   The motor of Problem 8.4 has a moment of inertia of 0.1 kg · m^2 and a load and friction torque of 65 N · m at 970 rpm. The motor is connected to a 60-Hz ac source through a diode, as in Fig. 8-11(a). The motor has a speed of 970 rpm when the conduction stops, at which moment the motor begins to coast. Determine the drop in the motor speed until the diode starts conducting again. Neglect the motor inductance.

The equation of motion during coasting is

$$J\dot{\omega}_m = b\omega_m = 0$$

which has the solution

$$\omega_m = \Omega_0 e^{-bt/J}$$

From the data, the numerical values are

$$\Omega_0 = 970 \text{ rpm} \qquad b = \frac{65}{970(2\pi)/60} = 0.64 \text{ N} \cdot \text{m} \cdot \text{s} \qquad J = 0.1 \text{ kg} \cdot \text{m}^2$$

so that

$$\omega_m = 970e^{-6.4t} \text{ (rpm)} \tag{1}$$

where t is in s. From Fig. 8-11(b), the coasting interval is of length

$$t_0 = \frac{2\pi - \gamma}{\omega} = \frac{2\pi - (2\pi/3)}{60(2\pi)} = \frac{1}{90} \text{ s}$$

where we have taken the value of γ from Problem 8.6(b). Then, from (1),

$$\omega_m(t_0) = 970e^{-6.4/90} = 903 \text{ rpm}$$

a decrease of $970 - 903 = 67$ rpm.

8.8.   The motor of Problems 8.4 and 8.7 is now operated by a thyristor at 60 Hz, as shown in Fig. 8-12, for which the firing angle α is 75° and the extinction angle β is 215°. Determine the drop in speed during coasting.

In this case, $\gamma = 215° - 75° = 140° = (7\pi/9)$ rad. Thus, the duration of coasting is

$$t_0 = \frac{2\pi - (7\pi/9)}{60(2\pi)} = 0.0102 \text{ s}$$

and, from (1) of Problem 8.7,

$$\omega_m(t_0) = 970e^{-(6.4)(0.0102)} = 908 \text{ rpm}$$

a drop of $970 - 908 = 62$ rpm.

8.9. The motor of Problems 8.4 and 8.7 is operated by a thyristor circuit at 110 V, 60 Hz (Fig. 8-7). If the firing angle is 60° and the extinction angle is 210°, (a) determine the average motor voltage over the conduction period. (b) If the motor develops an average torque of 4.8 N · m, what is the average armature current over the conduction period? (c) Find the average speed of the motor over the conduction period.

(a)   From (8.15), with γ = 210° − 60° = 150° = (5π/6) rad,

$$V_m' = \frac{110\sqrt{2}\ (\cos 60° - \cos 210°)}{5\pi/6} = 81.16 \text{ V}$$

(b)   From Problem 8.4,

$$k = \frac{0.08}{2\pi/60} = \frac{2.4}{\pi} \text{ V} \cdot \text{s} = \frac{2.4}{\pi} \text{ N} \cdot \text{m/A}$$

Then, from (8.17),

$$I' = \frac{2\pi}{k\gamma} T_{avg} = \frac{2\pi}{(2.4/\pi)(5\pi/6)} (4.8) = 15 \text{ A}$$

(c)   Since $E' = k\Omega'$, (8.15) gives

$$\Omega' = \frac{V_m' - RI'}{k} = \frac{81.16 - (0.51)(15)}{2.4/\pi} = 30.63\ \pi \text{ rad/s} = 919 \text{ rpm}$$

8.10. A dc motor is driven by a chopper as shown in Fig. 8-13(a). The data pertaining to the motor are: $R = 13$ mΩ, $L = 0.148$ mH, $b = 1.074 \times 10^{-6}$ N · m · s/rad. The torque constant of the motor is $k = 0.04$ N · m/A (where k is defined by $T_e = ki$, i being the armature current). For α = 0.5, λ = 33 ms, and $V_t = 48$ V, find the steady-state motor speed that marks the transition from Mode I to Mode II operation. What is the corresponding torque?

From the data,

$$\frac{\tau}{\lambda} = \frac{L}{R\lambda} = \frac{0.148}{13(33 \times 10^{-3})} = 0.345 \qquad \alpha = 0.5$$

Substituting these values into the equation of the separating curve,

$$1 = \frac{\tau}{\lambda} \ln\left(1 + \frac{3^{\alpha\lambda/\tau} - 1}{\Omega^*}\right)$$

we obtain

$$1 = 0.345 \ln\left(1 + \frac{e^{0.5/0.345} - 1}{\Omega^*}\right) \quad \text{or} \quad \Omega^* = \frac{1}{e^{0.5/0.345} + 1} = 0.19$$

and

$$\Omega_m = \frac{V_t\Omega^*}{k} = \frac{(48)(0.19)}{0.04} = 228 \text{ rad/s} = 2180 \text{ rpm}$$

The corresponding normalized torque is obtained from (8.23), using either expression on the right:

$$T^* = 0.5 - \left[\frac{(1.074 \times 10^{-6})(13 \times 10^{-3})}{(0.04)^2} + 1\right](0.19) = 0.31$$

Hence

$$T_{0\text{avg}} = \frac{kV_tT^*}{R} = \frac{(0.04)(48)(0.31)}{13 \times 10^{-3}} = 45.8 \text{ N} \cdot \text{m}$$

8.11. Consider a chopper-driven, dc series motor, as shown in Fig. 8-24(a). The motor chopper frequency is given in Fig. 8-24(b); the steady-state armature current is expected to be of the form given in Fig. 8-24(c). The motor saturation characteristic is shown in Fig. 8-25. The data pertaining to the motor and chopper are $R = 7.6$ mΩ, $L = 0.03$ mH, $b = 1.074 \times 10^{-6}$ N $\cdot$ m $\cdot$ s/A, k_a [see (4.6)] is 8, $\lambda = 1.0$ s, $\alpha = 0.5$, and $V_t = 24$ V. Obtain the motor torque-speed and developed power-speed characteristics.

The computations are too tedious unless a digital computer is employed. The steps are as follows:

(i) Store the saturation curve, ϕ versus I_{avg} in the computer. It suffices to use the three-segment approximation shown in Fig. 8-25.

(ii) Integration of the circuit equation over a period λ gives the average motor speed as

$$\Omega_m = \frac{\alpha V_t - I_{\text{avg}}R}{k_a\phi} \tag{1}$$

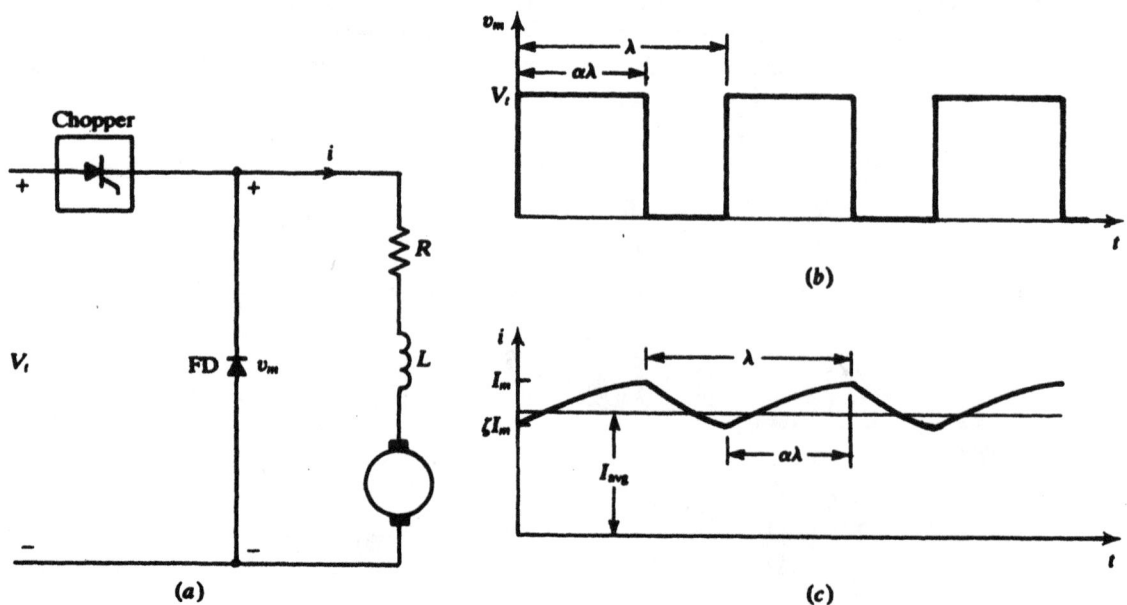

Fig. 8-24

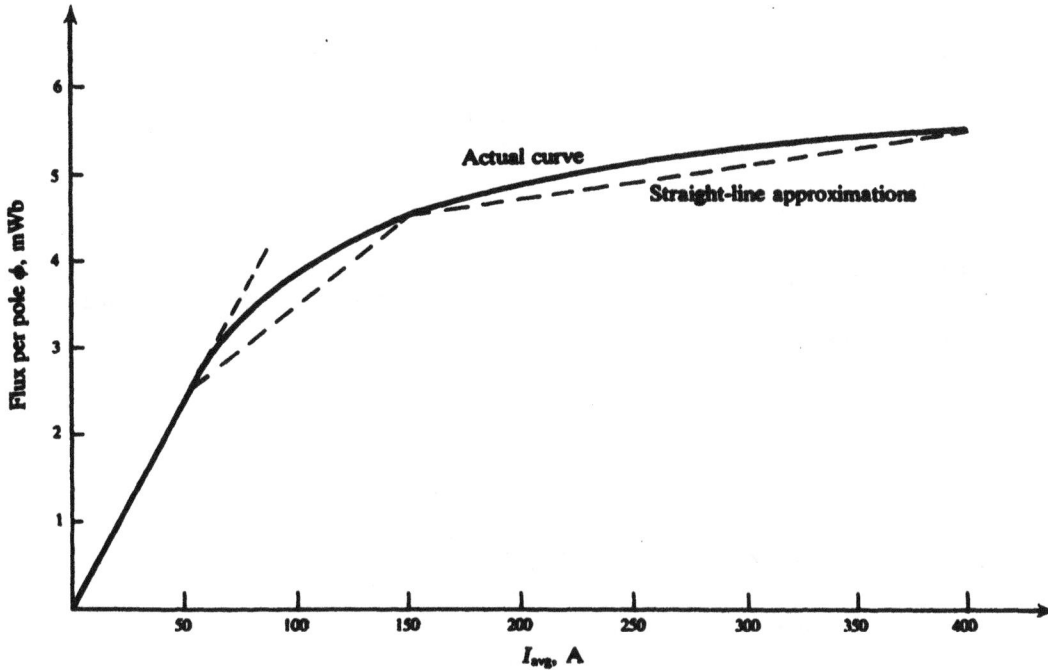

Fig. 8-25

Setting $\Omega_m = 10$ rad/s, iteratively choose $[I_{avg}, \phi(I_{avg})]$-pairs from (i) until (1) is satisfied; the final I_{avg} is the average armature current for the speed 10 rad/s.

(iii) Repeat (ii) for a set of evenly spaced values of Ω_m and thus obtain $I_{avg}(\Omega_m)$.

(iv) For each Ω_m, compute the corresponding developed power from

$$P_d = \alpha V_I I_{avg}(\Omega_m) - R[I_{avg}(\Omega_m)]^2$$

thus obtaining the power-speed characteristic.

(v) For each Ω_m, compute, from (iv),

$$T_{avg} = \frac{P_d}{\Omega_m}$$

generating the torque-speed characteristic.

8.12. The data relating to the single-phase inverter of Fig. 8-15(a) are as follows: $V = 48$ V, $R = 1$ Ω, $L = 5$ mH. For a fundamental frequency of 50 Hz, calculate the fundamental component of power supplied to the load by either source. The gating signals are as shown in Fig. 8-15(b).

$$\omega = 2\pi f = 2\pi(50) = 314 \text{ rad/s}$$

$$Z_1 = \sqrt{1^2 + (314 \times 0.005)^2} = 1.86 \text{ }\Omega$$

Then, from (8.26),

$$\text{rms fundamental current} = \frac{2(48)}{\sqrt{2}\,\pi(1.86)} = 11.6 \text{ A}$$

$$\text{power delivered to the load} = (11.6)^2(1) = 134.56 \text{ W}$$

$$\text{power from either source} = \frac{1}{2}(134.56) = 67.28 \text{ W}$$

8.13. The load on the inverter of Fig. 8-15(a) is changed to $R = 1\ \Omega$, $\omega L = 6\ \Omega$, and $1/\omega C = 8\ \Omega$, where the elements are all connected in series. If $V = 48$ V and $T = 3$ ms, (a) determine the average power supplied to the load. (b) Is load commutation possible? Consider only the fundamental components of current and voltage.

(a)   As in Problem 8.12,

$$I_1 = \frac{2(48)}{\sqrt{2}\,\pi\sqrt{1^2 + (6-8)^2}} = 9.66 \text{ A}$$

$$\text{output power} = (9.66)^2(1) = 93.4 \text{ W}$$

(b)   The phase angle of the current is given by

$$\tan \phi_1 = \frac{6-8}{1} = -2 \qquad \text{or} \qquad \phi_1 = -0.35\ \pi \text{ rad}$$

i.e. the current leads the voltage by $0.35\ \pi$. Therefore, the time available for turnoff is

$$\frac{0.35\pi}{\omega} = (0.35\pi)\left(\frac{T}{2\pi}\right) = 0.175\ T = (0.175)(3 \times 10^{-3}) = 525\ \mu s$$

which is much greater than a typical thyristor turnoff time ($\sim$100 μs). Hence load commutation is possible.

8.14. An R-L circuit is fed from an ac source in series with a thyristor, as shown in Fig. 8-26(a). The thyristor is fired at an angle α. Derive the condition for line commutation, and determine the thyristor conduction period.

The voltage equation is

$$L\frac{di}{dt} + Ri = V_m \sin \omega t \qquad (1)$$

where the symbols are defined in Fig. 8-26(a). The solution to (1) is of the form (See Section 8.4)

$$i = \frac{V_m}{Z} \sin(\omega t - \phi) + ke^{-(R/L)t} \qquad (2)$$

where $Z = \sqrt{R^2 + (\omega L)^2}$ and $\tan \phi = \omega L/R$. To evaluate k, we use the condition $i = 0$ at $\omega t = \alpha$ in (2) to obtain

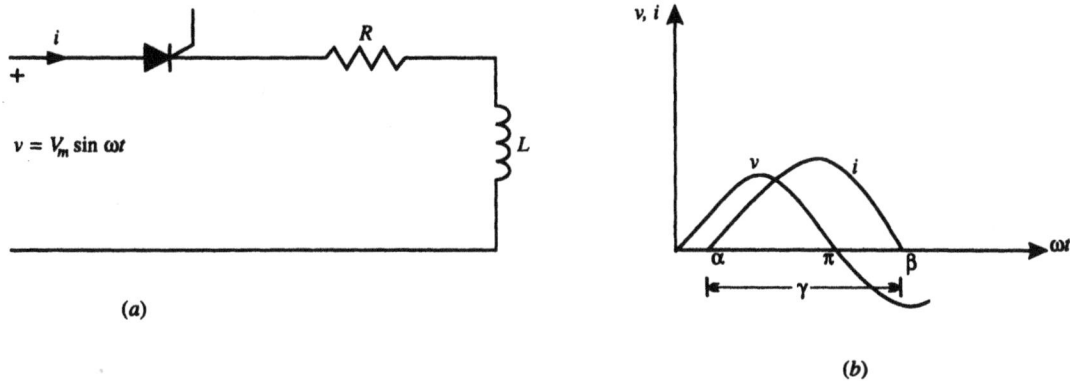

(a)

(b)

Fig. 8-26

$$i = \frac{V_m}{2} \left[\sin(\omega t - \phi) - \sin(\alpha - \phi)e^{R(\alpha - \beta)/\omega L} \right] \tag{3}$$

To obtain the condition for line commutation we notice that $i = 0$ at $\omega t = \beta$, as shown in Fig. 8-26(b). Thus the required condition for commutation is given by the transcendental equation

$$\sin(\beta - \phi) = \sin(\alpha - \phi)e^{R(\alpha - \beta)/\omega L}$$

The conduction angle γ is then given by

$$\gamma = \beta - \alpha \tag{5}$$

8.15. From Problem 8.14 it is clear that line commutation is possible only in ac systems. Similarly, in a dc system, load commutation is not possible for an *R-L* circuit. However, load commutation in an *R-L-C* circuit fed from a dc source is possible if the circuit current could be made oscillatory. For the circuit shown in Fig. 8-27, derive the equation governing the time of commutation.

The voltage equation for the given circuit is

$$L \frac{di}{dt} + Ri + \frac{1}{C} \int_0^t i \, dt + v_c(0) = V \tag{1}$$

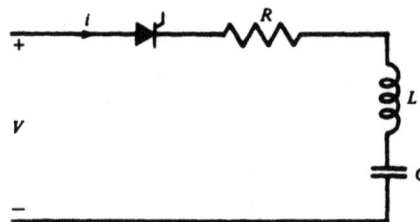

Fig. 8-27

where $v_c(0)$ is the charge on the capacitor at $t = 0$. Defining

$$\zeta = \frac{R}{2L} = \text{damping ratio} \tag{2}$$

$$\omega_0 = \frac{1}{\sqrt{LC}} = \text{resonant frequency} \tag{3}$$

for $\zeta < \omega_0$, the solution to (1) becomes

$$i(t) = e^{\zeta t}(A \cos \omega_r t + B \sin \omega_r t) \tag{4}$$

where A and B are arbitrary constants and

$$\omega_r = \sqrt{\omega_0^2 - \zeta^2} = \text{ringing frequency} \tag{5}$$

Because of the inductance, the current in the circuit cannot change instantaneously. Hence $i(0) = 0$, which when substituted in (4) yields $A = 0$, and (4) then becomes

$$i(t) = Be^{-\zeta t} \sin \omega_r t \tag{6}$$

for commutation, this current must go zero, requiring that

$$\omega_r t = \pi \tag{7}$$

8.16. For the circuit shown in Fig. 8-27, $V = 96$ V, $L = 50$ mH, $C = 80$ μF, and $R = 40$ Ω. The initial charge on the capacitor is zero. At what time will the thyristor turn off?

From (2) and (3) of Problem 8.15 we have

$$\zeta = \frac{40}{2 \times 50 \times 10^{-3}} = 400$$

and

$$\omega_0 = \frac{1}{\sqrt{50 \times 10^{-3} \times 80 \times 10^{-6}}} = 500 \text{ rad/s}$$

From (5) of Problem 8.15

$$\omega_r = \sqrt{500^2 - 400^2} = 300 \text{ rad/s}$$

Finally, (7) of Problem 8.14 yields

$$300t = \pi$$

or

$$t = \frac{\pi}{300} = 10.47 \text{ ms}$$

8.17. Analyze the forced-commutation circuit shown in Fig. 8-20, for the underdamped case. In particular, (a) determine i_1, subject to the initial conduction $i_1(0) = 0$; (b) obtain an expression for the voltage across the capacitor. (c) What is the width of each current pulse? (d) At what time does the current pulse reach its maximum?   (e) What is the maximum current?

(a)   When the thyristor T_1 is turned on, the current in the RLC-series circuit satisfies

$$Ri_1 + L \frac{di_1}{dt} + \frac{1}{C} \int i_1 \, dt = V \tag{1}$$

For underdamping, $4L/C > R^2$, and (1) is solved by standard methods to give

$$i_1 = \frac{V_0}{Z_0} e^{-\alpha t} \sin \beta t \tag{2}$$

where $V_0 \equiv V - V_{c0}$

$V_{c0} \equiv$ voltage across C at $t = 0$ (an arbitrary constant)

$Z_0 \equiv \sqrt{\frac{L}{C} - \frac{R^2}{4}}$ = characteristic impedance

$\alpha \equiv \frac{R}{2L}$ = attenuation constant

$\beta \equiv \frac{Z_0}{L}$ = phase constant

(b)   The voltage across the capacitor, v_c, is given by

$$v_c = V - Ri_1 - L \frac{di_1}{dt} = V - \frac{V - V_{c0}}{\sin \varepsilon} e^{-\alpha t} \sin (\beta t + \varepsilon)$$

where $\tan \varepsilon = \beta/\alpha$.

(c)   pulse width $= \dfrac{\pi}{\beta}$

(d)   By differentiation of (2), we find

$$t_{max} = \frac{\varepsilon}{\beta} \tag{3}$$

(e)   From (2) and (3),

$$I_{max} = \frac{V_0}{Z_0} e^{-\alpha/\beta} \sin \varepsilon$$

8.18.   For the circuit of Fig. 8-20, $V = 96$ V, $R = 0.2\ \Omega$, $L = 0.05$ mH, and $C = 10\ \mu$F. Determine the current pulse width, and the time when the current pulse is maximum immediately after the SCR is turned on. What is the value of the maximum current, if the capacitor is uncharged when the SCR is turned on?

From the results of Problem 8.17,

$$Z_0 = \sqrt{\frac{5 \times 10^{-5}}{10 \times 10^{-6}} - \frac{(0.2)^2}{4}} = 2.23 \ \Omega$$

$$\beta = \frac{2.23}{5 \times 10^{-3}} = 4.46 \times 10^4 \ \text{rad}/s$$

$$\text{pulse width} = \frac{\pi}{4.46 \times 10^4} = 70.3 \ \mu s$$

$$\alpha = \frac{0.2}{2(5 \times 10^{-5})} = 2 \times 10^3 \ s^{-1}$$

$$\epsilon = \tan^{-1}\left(\frac{4.46 \times 10^4}{2 \times 10^3}\right) = 87.4° = 1.53 \ \text{rad}$$

$$t_{max} = \frac{1.53}{4.46 \times 10^4} = 34.2 \ \mu s$$

$$V_0 = 96 \ \text{V}$$

$$I_{max} = \frac{96}{2.23} \ e^{-(2\times10^3)(34.2\times10^{-6})} \sin 87.4° = 40.2 \ \text{A}$$

8.19. Determine the voltage across C in Problem 8.18 at the end of the current pulse.

From the results of Problem 8.17(b) and (c),

$$v_c = V(1 + e^{-\alpha\pi/\beta})$$

$$= 96\left[1 + e^{-(2\times10^3)(70.3\times10^{-6})}\right] = 179.4 \ \text{V}$$

8.20. The current of a chopper-controlled motor with a freewheeling diode is plotted in Fig. 8-24(c); approximate the curve by straight-line segments. The maximum current is 60 A and the minimum current is 40 A. The period λ is 10 ms and the duty cycle α is 0.6. Determine the rms and the average values of (a) the input current, (b) the freewheeling-diode current, and (c) the load current.

(a)   The input current has a waveform similar to that of Fig. 8-21, for which Problem 8.1 gives

$$A = A_m\left[\frac{T_o}{T_p} \frac{1}{3} (1 + \zeta + \zeta^2)\right]^{1/2}$$

In the present case,

$$A_m = 60 \ \text{A} \qquad \zeta = \frac{40}{60} = \frac{2}{3} \qquad \frac{T_o}{T_p} = \alpha = 0.6$$

and so

$$I = 60 \sqrt{0.6 \times \frac{1}{3}\left(1 + \frac{2}{3} + \frac{4}{9}\right)} = 39 \text{ A}$$

Again, from Problem 8.1,

$$A_{avg} = \frac{T_o}{2T_p} A_m (1 + \zeta)$$

or

$$I_{avg} = \frac{0.6}{2}(60)\left(1 + \frac{2}{3}\right) = 30 \text{ A}$$

(b)   The freewheeling-diode current also has the waveform of Fig. 8-21, except that in this case

$$\frac{T_o}{T_p} = 1 - \alpha = 0.4$$

other values remaining the same as in (a) above. Thus

$$I = 60 \sqrt{0.4 \times \frac{1}{3}\left(1 + \frac{2}{3} + \frac{4}{9}\right)} = 31.8 \text{ A}$$

$$I_{avg} = \frac{0.4}{2}(60)\left(1 + \frac{2}{3}\right) = 20 \text{ A}$$

(c)   For the linearized current waveform of Fig. 8-24(c), it may be shown as in Problem 8.1 that

$$I = I_m \sqrt{\frac{1}{3}(1 + \zeta + \zeta^2)} = 60 \sqrt{\frac{1}{3}\left(1 + \frac{2}{3} + \frac{4}{9}\right)} = 50.3 \text{ A}$$

$$I_{avg} = \frac{1}{2}I_m(1 + \zeta) = \frac{60}{2}\left(1 + \frac{2}{3}\right) = 50 \text{ A}$$

8.21.   A chopper-driven, separately excited, dc motor has an armature resistance of 0.1 Ω and an inductance of 0.25 mH. The back emf constant [see (4.8)] at rated field current of 7.64 A, is 0.05 V · s/A, and the chopper input voltage is $V_t = 96$ V. The chopper has a freewheeling diode and the motor armature current is controlled by the on-time of the current pulse. For a 500-Hz pulse frequency, obtain the range of on-time pulse width to control the armature current from 10 to 200 A at 2000 rpm. For large average current, assume the current waveform to be given by the piecewise-linear curve of Fig. 8-28(a).

$$\text{motor speed} = \frac{2\pi(2000)}{60} = 209.4 \text{ rad/s}$$

$$\text{motor back emf} \equiv V_0 = k\omega_m I_f = (0.05)(209.4)(7.64) = 80 \text{ V}$$

If the motor resistance is neglected, then during the off-time period the armature current i flows through the freewheeling diode and obeys

$$L\frac{di}{dt} + V_0 = 0 \quad \text{or} \quad \frac{di}{dt} = -\frac{V_0}{L}$$

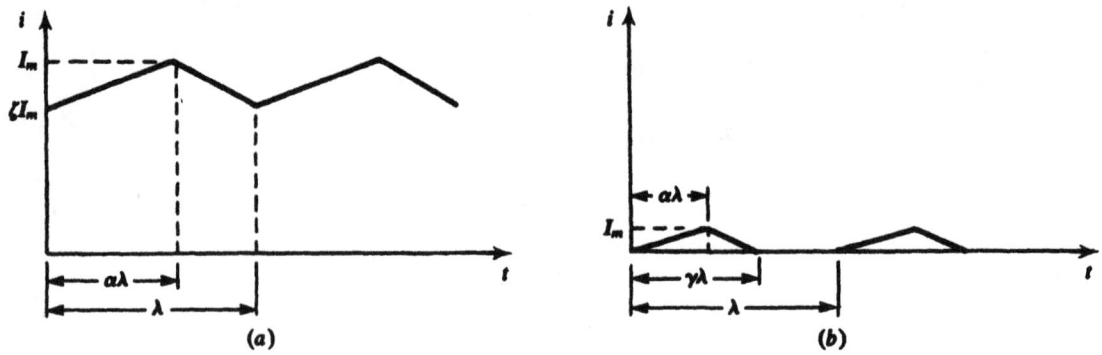

Fig. 8-28

Thus, the slope of the descending portions of the curve in Fig. 8-28(a) is $-V_0/L$, and we have from the figure

$$\frac{(1 - \zeta)I_m}{(\alpha - 1)\lambda} = -\frac{V_0}{L} \tag{1}$$

On the other hand, during the on-time,

$$L\frac{di}{dt} + V_0 = V_t \qquad \text{or} \qquad \frac{di}{dt} = \frac{V_t - V_0}{L}$$

and so

$$\frac{(1 - \zeta)I_m}{\alpha\lambda} = \frac{V_t - V_0}{L} \tag{2}$$

Dividing (2) and (1), we obtain

$$\alpha = \frac{V_0}{V_t} = \frac{80}{96}$$

Now, corresponding to 500 Hz,

$$\lambda = \frac{1}{500} \text{ s} = 2 \text{ ms}$$

and

$$\alpha\lambda = \frac{80}{96} (2) = 1.67 \text{ ms}$$

This represents the upper limit of the desired range.

For low current values, $i \approx 10$ A, assume the waveform shown in Fig. 8-28(b). As above, we infer that

$$\frac{\alpha}{\gamma} = \frac{V_0}{V_t} = \frac{80}{96} \tag{3}$$

Moreover,

$$I_{avg} = \frac{1}{\lambda}\left(\frac{1}{2}\gamma\lambda I_m\right) = \frac{1}{2}\gamma I_m \tag{4}$$

and

$$\frac{I_m}{\alpha\lambda} = \frac{V_t - V_0}{L} \tag{5}$$

Elimination of γ and I_m from (3), (4), and (5) yields

$$\alpha\lambda = \left[\frac{2I_{avg}\,L\lambda}{(V_t - V_0)(V_t/V_0)}\right]^{1/2} \tag{6}$$

In (6) substitute $L = 0.25 \times 10^{-3}$ H, $V_t = 96$ V, $V_0 = 80$ V, $\lambda = 2 \times 10^{-3}$ s, and $I_{avg} = 10$ A to get

$$\alpha\lambda = \left[\frac{2(10)(0.25 \times 10^{-3})(2 \times 10^{-3})}{(96 - 80)(96/80)}\right]^{1/2} = 0.72 \text{ ms}$$

which is the lower limit of the range of on-time.

Supplementary Problems

8.22. Determine the rms and average values of the current waveforms shown in Fig. 8-29.
 Ans. (a) $I_m(T_d/T_p)^{1/2}$, $I_m(T_d/T_p)$; (b) $I_m(T_d/3T_p)^{1/2}$, $I_m(T_d/2T_p)$;

 (c) $\dfrac{I_m}{\sqrt{2}\,T_o}\left[\dfrac{1}{2}\,(T_o - T_\alpha) + \dfrac{T_o}{4\pi}\,\sin\,(2\pi T_\alpha/T_o)\right]^{1/2}$, $\dfrac{I_m T_o}{\pi T_p}\left[1 + \cos\,(\pi T_\alpha/T_o)\right]$

8.23. Show that (8.5) is a valid solution to the differential equation

$$L\,\frac{di}{dt} + ri = v_o$$

for the rectifier circuit of Fig. 8-5(a).

8.24. Explain why no average value of the rectified voltage can appear across the inductor in Fig. 8-5(a).

8.25. An electronic motor controller has a chopped half-wave rectified sinusoidal output voltage waveform, as shown in Fig. 8-30. Determine the average and rms values of the output voltage.

 Ans. $\dfrac{2V_m}{\pi}\left(\dfrac{T_0}{T_p}\right)$; $V_m\sqrt{\dfrac{T_0}{2T_p}}$

8.26. A half-wave rectifier for battery charging may be modeled by the circuit shown in Fig. 8-31. Given values are $v_s = 14 \sin 100t$, $V_0 = 10$ V, and $R = 0.2\ \Omega$.
 (a)   Sketch v_0.
 (b)   Find an expression for $i_o(t)$ for $0 < \omega t \le T$.
 (c)   Calculate the average and rms values of i_o.
 (d)   Determine the power delivered to the battery.

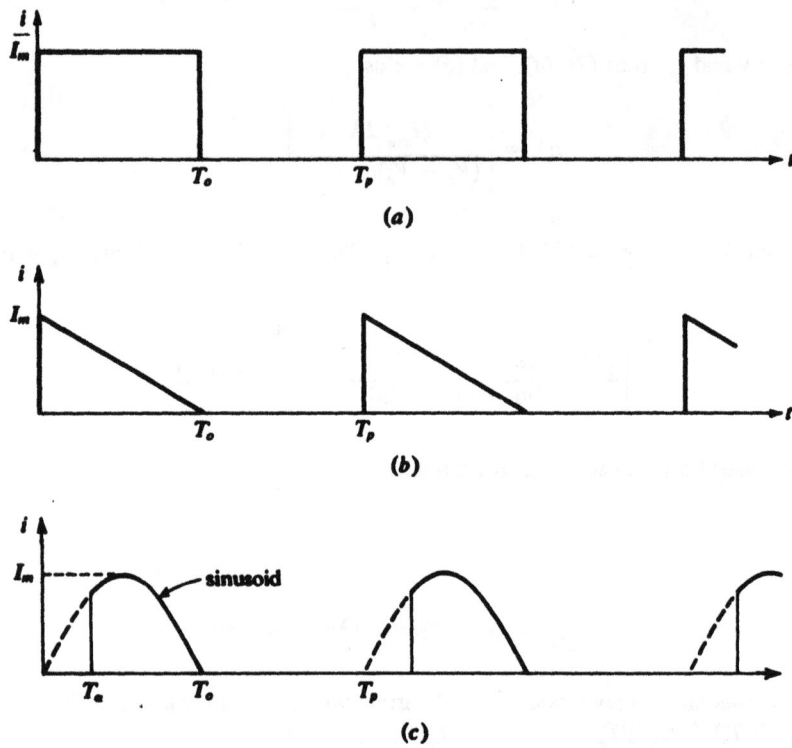

(a)

(b)

(c)

Fig. 8-29

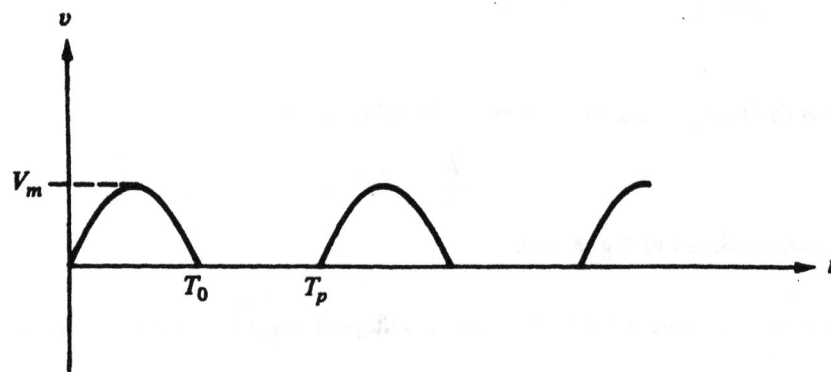

Fig. 8-30

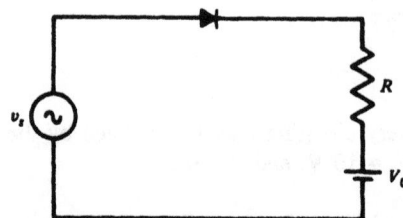

Fig. 8-31

Ans. (*a*) Diode conduits from α to β with α = 45.6° and β = 134.4°; (*b*) $i_o(t) = 70 \sin \omega t$, $\alpha \le \omega t \le \beta$, otherwise zero; (*c*) 7.2 A; (*d*) 42.94 W (average)

8.27. An ac source, having an rms voltage V, feeds a purely resistive load through an SCR. If the firing angle of the SCR is α, what is the average voltage across the load? *Ans.* $0.225V(1 + \cos \alpha)$

8.28. Determine the magnitudes of the average or dc voltages in the following rectifier circuits; (*a*) single-phase, half-wave; (*b*) single-phase, full-wave; (*c*) 3-phase, half-wave; (*d*) 3-phase, full-wave. Assume the same maximum value and the same frequency of the input voltage wave in all cases.
Ans. (*a*) V_m/π; (*b*) $2V_m/\pi$; (*c*) $(3\sqrt{3})V_m/2\pi$; (*d*) $3V_m/\pi$

8.29. An impedance of $(3 + j4)\ \Omega$ is fed by a 220-V, 60-Hz ac source through a single diode. What is the average value of the current? *Ans.* 25.97 A

8.30. A dc motor is supplied by a 200-V, 60-Hz, half-wave rectifier and operates at a constant speed of 1100 rpm. The motor parameters are: armature resistance, 0.1 Ω; armature inductance, 0.46 mH; moment of inertia, 0.9 kg · m²; friction coefficient, 4.5 N · m · s/rad; back emf constant, 0.141 V/rpm. Calculate the average values of (*a*) the armature current, (*b*) the torque, and (*c*) the developed power.
Ans. (*a*) 35 A; (*b*) 47 N · m; (*c*) 5.43 kW

8.31. In Problem 8.30, neglect the armature-circuit inductance. Determine when the diode starts to conduct.
Ans. α = 30°

8.32. The motor of Problem 8.30 is running at 1100 rpm when the diode ceases to conduct; and until the next conduction, the motor coasts. Neglecting the armature inductance, compute the percent drop in speed from the initial speed. *Ans.* 12.4%

8.33. The motor of Problem 8.30 is operated by a 220-V, 60-Hz thyristor for which the firing angle is 160° and the extinction angle is 190°. Determine (*a*) the average armature current and (*b*) the average motor speed.
Ans. (*a*) 262 A; (*b*) 749 rpm

8.34. Figure 8-32 shows a dc motor that is energized and controlled through an SCR from a square-wave ac source. The square-wave input voltage has a magnitude of 100 V and a frequency of 100 Hz. The armature inductance is 1 mH. The motor has a freewheeling diode connected as shown. The motor is operating at a constant speed such that its back emf, V_o, is 50 V, and the load is of so large an inertia that this emf can be assumed constant during a complete cycle of the voltage wave. Also, the armature resistance and the freewheeling-diode voltage drop during diode conduction are negligible. If SCR turn-on is delayed for 60° after the start of the positive half-cycle of input voltage, (*a*) what is the magnitude of the armature current when the SCR turns off and the diode starts to conduct? (*b*) How much energy is stored in the inductance at this instant? (*c*) Will the current in the freewheeling loop around the armature reach zero magnitude before the SCR turns on again at the 60° point of the next positive half-cycle of the voltage? (*d*) What is the average armature current? (*e*) What is the voltage across the SCR (anode-cathode) an instant after it turns off?
Ans. (*a*) 167 A; (*b*) 13.94 J; (*c*) yes; (*d*) 111 A; (*e*) −100 V

8.35. Verify (*8.27*).

8.36. For the inverter of Fig. 8-15(*a*), $V = 36$ V, $R = 2.5\ \Omega$, and $\omega L = 2.5\ \Omega$ (at the fundamental frequency). Determine the power supplied to the load. By calculating the source current, show that the power is equally shared by the two sources. *Ans.* 52.5 W

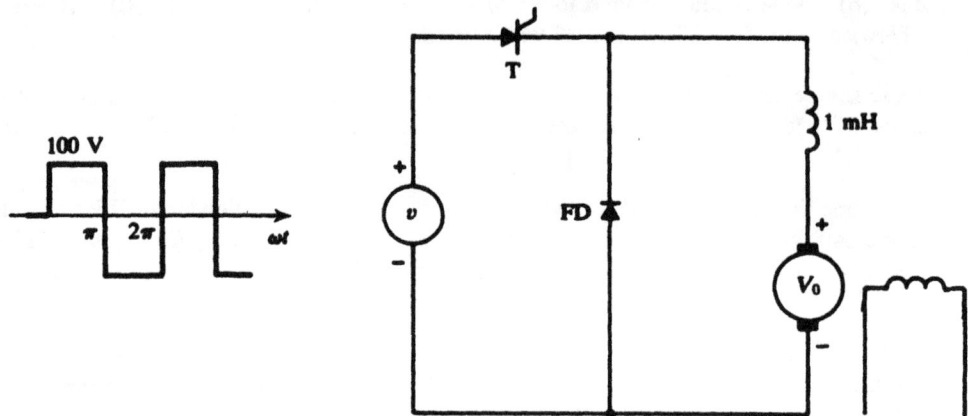

Fig. 8-32

8.37. A capacitive reactance of 4.0 Ω is added in series with the load given in Problem 8.36. For an inverter frequency of 250 Hz, determine if forced commutation is necessary, by computing the time available for turnoff.   *Ans.* Not necessary

8.38. A chopper has a frequency of 200 Hz and a duty cycle of 0.4. It controls a dc motor and has a freewheeling diode. The motor current fluctuates between 36 A and 18 A. Calculate the rms and average values of (a) the diode current and (b) the motor current.
       Ans. (a) 17.44 A, 10.8 A; (b) 27.5 A, 27 A

8.39. A chopper-driven dc motor has 0.1 mH armature inductance and 10 mΩ armature resistance. The chopper operates at 72 V and 200 Hz, and has a freewheeling diode. If the motor runs at 1500 rpm, has a back emf constant of 0.4 V/rpm at the operating field current, and never has zero armature current, determine (a) the on-time pulse width and (b) the duty cycle. Neglect the effect of armature resistance.
       Ans. (a) 4.17 ms; (b) 0.83

Appendix A

Units Conversion

Symbol	Description	One: (SI Unit)	Is Equal to: (English Unit)	(CGS Unit)
B	Magnetic flux density	$T \, (= 1 \text{ Wb/m}^2)$	$6.452 \times 10^4 \text{ lines/in}^2$	10^4 gauss
H	Magnetic field intensity	A/m	0.0254 A/in	$0.004\pi \times 10^{-3}$ oersted
ϕ	Magnetic flux	Wb	10^8 lines	10^8 maxwells
b	Viscous damping coefficient	$N \cdot m \cdot s$	0.73756 lb-ft-s	10^7 dyne-cm- s
F	Force	N	0.2248 lb	10^5 dynes
J	Moment of inertia	$kg \cdot m^2$	23.73 lb-ft^2	10^7 g-cm^2
T	Torque	$N \cdot m$	0.73756 ft-lb	10^7 dyne-cm
W	Energy	J	1 Watt-s	10^7 ergs

Note: lbf = pound force; lbm = pound mass

Appendix B

Characteristics of Single-Film-Coated, Rounded, Magnet Wire

AWG Size	Bare Wire Diameter (Nominal), in	Film Additions, in		Overall Diameter, in			Weight at 20°C (68°F)		Resistance at 20°C (68°F)		Wires/in (Nom.)	AWG Size
		Min.	Max.	Min.	Nom.	Max.	Lbs/1000 ft (Nom.)	Ft/lb (Nom.)	Ohms/1000 ft (Nom.)	Ohms/lb (Nom.)		
8	.1285	.0016	.0026	.1288	.1306	.1324	50.20	19.92	.6281	.01251	7.66	8
9	.1144	.0016	.0026	.1149	.1165	.1181	39.81	25.12	.7925	.01991	8.58	9
10	.1019	.0015	.0025	.1024	.1039	.1054	31.59	31.66	.9988	.03162	9.62	10
11	.0907	.0015	.0025	.0913	.0927	.0941	25.04	39.94	1.26	.05032	10.8	11
12	.0808	.0014	.0024	.0814	.0827	.0840	19.92	50.20	1.59	0.7982	12.1	12
13	.0720	.0014	.0023	.0727	.0728	.0750	15.81	63.25	2.00	.1265	13.5	13
14	.0641	.0014	.0023	.0649	.0659	.0670	12.49	80.06	2.52	.2018	15.2	14
15	.0571	.0013	.0022	.0578	.0588	.0599	9.948	100.5	3.18	.3196	17.0	15
16	.0508	.0012	.0021	.0515	.0525	.0534	7.880	126.9	4.02	.5101	19.0	16
17	.0453	.0012	.0020	.0460	.0469	.0478	6.269	159.5	5.05	.8055	21.3	17
18	.0403	.0011	.0019	.0410	.0418	.0426	4.970	201.2	6.39	1.286	23.9	18
19	.0359	.0011	.0019	.0366	.0374	.0382	3.943	253.6	8.05	2.041	26.7	19
20	.0320	.0010	.0018	.0327	.0334	.0341	3.138	318.7	10.1	3.219	29.9	20
21	.0285	.0010	.0018	.0292	.0299	.0306	2.492	401.2	12.8	5.135	33.4	21
22	.0253	.0010	.0017	.0260	.0267	.0273	1.969	507.9	16.2	8.228	37.5	22
23	.0226	.0009	.0016	.0233	.0238	.0244	1.572	636.1	20.3	12.91	42.0	23
24	.0201	.0009	.0015	.0208	.0213	.0218	1.240	806.5	25.7	20.73	46.9	24
25	.0179	.0009	.0014	.0186	.0191	.0195	.988	1012	32.4	32.79	52.4	25
26	.0159	.0008	.0013	.0165	.0169	.0174	.779	1284	41.0	52.64	59.2	26
27	.0142	.0008	.0013	.0149	.0153	.0156	.623	1605	51.4	82.50	65.4	27
28	.0126	.0007	.0012	.0132	.0136	.0139	.491	2037	65.3	133.0	73.5	28
29	.0113	.0007	.0012	.0119	.0122	.0126	.395	2532	81.2	205.6	82.0	29
30	.0100	.0006	.0011	.0105	.0109	.0112	.310	3226	104	335.5	91.7	30
31	.0089	.0006	.0011	.0094	.0097	.0100	.246	4065	131	532.5	103	31
32	.0080	.0006	.0010	.0085	.0088	.0091	.199	5025	162	814.1	114	32
33	.0071	.0005	.0009	.0075	.0078	.0081	.157	6394	206	1317	128	33
34	.0063	.0005	.0008	.0067	.0070	.0072	.123	8130	261	2122	143	34
35	.0056	.0004	.0007	.0059	.0062	.0064	.0977	10235	331	3388	161	35
36	.0050	.0004	.0007	.0053	.0056	.0058	.0783	12771	415	5300	179	36
37	.0045	.0003	.0006	.0047	.0050	.0052	.0632	15823	512	8101	200	37
38	.0040	.0003	.0006	.0042	.0045	.0047	.0501	19960	648	12934	222	38
39	.0035	.0002	.0005	.0036	.0039	.0041	.0383	26110	847	22115	256	39
40	.0031	.0002	.0005	.0032	.0035	.0037	.0301	33222	1080	35880	286	40
41	.0028	.0002	.0004	.0029	.0031	.0033	.0244	40984	1320	54099	323	41
42	.0025	.0002	.0004	.0026	.0028	.0030	.0195	51282	1660	85128	357	42
43	.0022	.0002	.0003	.0023	.0025	.0026	.0153	65360	2140	139870	400	43
44	.0020	.0001	.0003	.0020	.0022	.0024	.0124	80645	2590	208870	455	44

Characteristics of Magnetic Materials
and Permanent Magnets

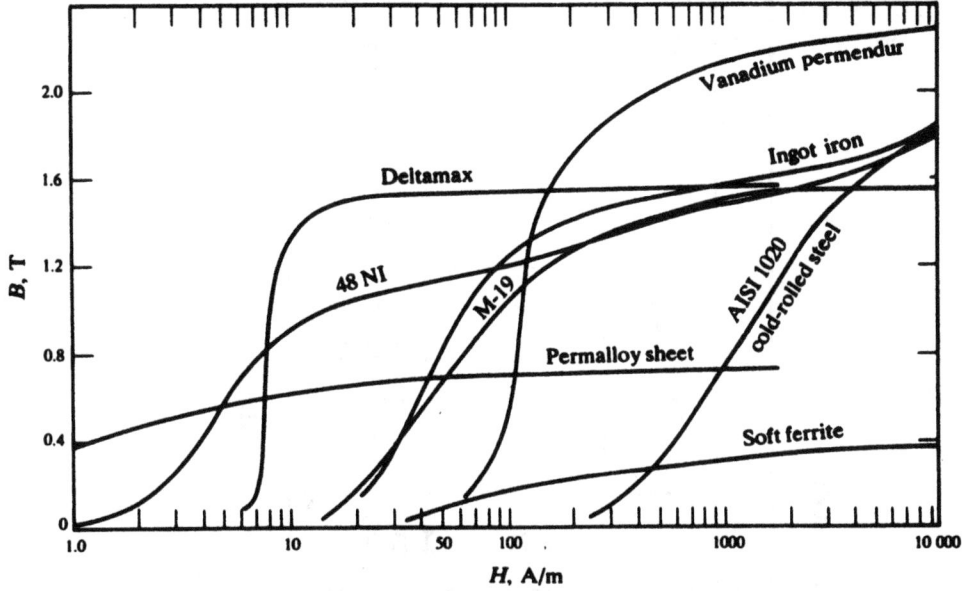

Fig. C-1. *B-H* curves of selected soft magnetic materials.

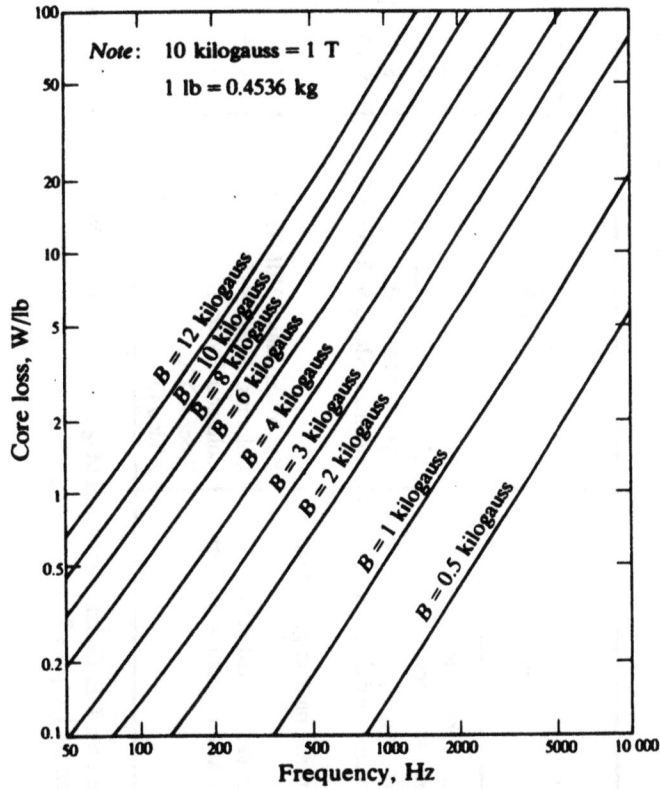

Fig. C-2. Core loss for nonoriented silicon steel, 0.019-in-thick lamination.

Table C-1. Properties of commercial cast alnicos*

Type	Brand Names	Percentage Composition (Balance, Fe)					Magnetic Properties†		
		Al	Ni	Co	Cu	Others	B_r, T	$(BH)_{max}$, kJ/m^3	H_c, kA/m
Isotropic, cobalt-free	Alni, Alnico III	12-14	24-26		0-4	0-1 Ti	0.5-0.6	10	57-40
Isotropic, with cobalt	Alnico, Alnico 2, Reco	9-11	16-20	12	3-6	0-1 Ti	0.65-0.8	13-14	50-60
Isotropic, high H_c	Hynico, ALnico XII	8-10	18-21	17-20	2-4	4-8 Ti	0.6-0.7	14-16	72-60
Field-treated random grain	Alcomax, Alnico V, Ticonal G	8-8.5	13-15	24	2-4	0-1 Ti 0-2 Nb	1.1-1.3	36-44	60-45
Field-treated random grain, high H_c	Hycomax, Alnico VIII, Ticonal X	7-8	14-15	34-40	3-4	4-8 Ti	0.75-0.90	40-48	160-110
Field-treated directed grain	Columax, Alnico V-7	8-8.5	13-15	24	2-4	0-1 Nb	1.3-1.4	56-64	62-56
Field-treated directed grain, high H_c	Columnar Hycomax, Ticonal XX, Alnico IX	7-8	14-16	24-40	3-4	4-6 Ti 0-1 Nb 0.3 S	1-1.1	50-75	140-110

*Adapted from J. E. Gould, "Permanent Magnets," *IEE Reviews*, Vol. 125, No. 11R, November 1978, pp. 1137–1151.
†See Example 1.1, page 7.

Table C-2. Properties of commercial ferrite magnets*

Type	Brand Names	Specific Gravity	Magnetic Properties†		
			B_r, T	$(BH)_{max}$, kJ/m^3	H_c, kA/m
Sintered isotropic	Magnadur 1, Feroba 1	4.8	0.22	8	140
Sintered anisotropic	Magnadur 2, Feroba 2	5.0	3.6	29	150
Sintered anisotropic, high H_c	Magnadur 3, Feroba 3	4.9	0.36	26	240
Plastic bonded		3.6	0.14	3	90
Plastic bonded, anisotropic		0.39	0.2	2	110

*Adapted from J. E. Gould, "Permanent Magnets," *IEE Reviews*, Vol. 125, No. 11R, November 1978, pp. 1137-1151.
†See Example 1.1, page 7.

Index